TECHNOHISTORY

Using the History of American Technology in Interdisciplinary Research

Edited by
Chris Hables Gray

OPEN FORUM SERIES

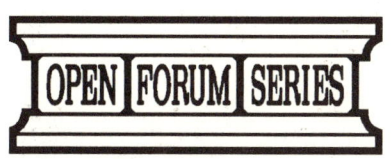

KRIEGER PUBLISHING COMPANY
MALABAR, FLORIDA
1996

Original Edition 1996

Printed and Published by
KRIEGER PUBLISHING COMPANY
KRIEGER DRIVE
MALABAR, FLORIDA 32950

Copyright © 1996 by Chris Hables Gray

All rights reserved. No part of this book may be reproduced in any form or by any means, electronic or mechanical, including information storage and retrieval systems without permission in writing from the publisher.
No liability is assumed with respect to the use of the information contained herein. Printed in the United States of America

---

FROM A DECLARATION OF PRINCIPLES JOINTLY ADOPTED BY A COMMITTEE OF THE AMERICAN BAR ASSOCIATION AND A COMMITTEE OF PUBLISHERS:

This publication is designed to provide accurate and authoritative information in regard to the subject matter covered. It is sold with the understanding that the publisher is not engaged in rendering legal, accounting, or other professional service. If legal advice or other expert assistance is required, the services of a competent professional person should be sought.

Library of Congress Cataloging-In-Publication Data
Technohistory : using the history of American technology in interdisciplinary research / Chris Hables Gray, editor.
   p. cm.
  Includes bibliographic references and index.
  ISBN 0-89464-853-5 (acid-free paper)
  1. Technology—United States—History—Case studies.
  2. Technology—United States—Historiography—Case studies.
I. Gray, Chris Hables.
T21.T38 1996
609–dc20   94-36684
                                                                       CIP

10 9 8 7 6 5 4 3 2

Dedicated to Patty L. Paulson (1952–1994)
and the Oregon State University Center for the Humanities
which she served with such grace

CONTENTS

Preface		vii
Acknowledgments		xiii
Contributors		xv
Introduction: Reclaiming Technology for the Humanities *Carroll Pursell, Case Western Reserve University*		1
1	Digital Color Imaging and the Colorization Controversy: Culture, Technology, and the Popular as Lightning Rod *Gary Edgerton, Old Dominion University*	5
2	The Waste Isolation Pilot Plant: Supertechnology in Limbo *Monika Ghattas, Technical Vocational Institute*	33
3	Nursing, Technology, and Gender: A History of Progress or Colonization? *Lena Sorensen, University of Massachusetts, Boston*	65
4	Promoting Critical Thinking Through Teaching Technohistory *Myra Jones, Manatee Community College*	81
5	Waterpower on the Sugar River in Newport, New Hampshire: A Historical and Industrial Archaeological Investigation *Walter Ryan, New Hampshire Technical College*	97
6	Electrical Communication, Language, and Self *David Hochfelder, Case Western Reserve University*	119
7	Medical Cyborgs: Artificial Organs and the Quest for the Posthuman *Chris Hables Gray, University of Great Falls*	141
8	Some Body Fantasies in Cyberspace Texts: A View from Its Exclusions *Heidi J. Figueroa-Sarriera, Río Piedras Campus, University of Puerto Rico*	179

9	Manifest(o) Technologies: Marx, Marinetti, Haraway *Steven Mentor, University of Washington*	195
10	NASA Retrospect and Prospect: Space Policy in the 1950s and the 1990s *Roger D. Launius, NASA History Division*	215
11	Color Matters: Race, History, and Florida's Sara Lee Doll *Gordon Patterson, Florida Institute of Technology*	233

Afterword: Rethinking Technohistory　　　　　　　　　251
　　Chris Hables Gray, University of Great Falls

Suggested Readings in Technohistory　　　　　　　　259

Index　　　　　　　　　　　　　　　　　　　　　　263

PREFACE
Chris Hables Gray

The goal of this book is not only to show that the history of technology is interesting and exciting, but that it is important as well—and not just important as its own academic discipline and to other academic disciplines, but in many other areas of life, from public policy to ethics. I believe that the history of technology is not just for historians talking to themselves. It is applicable to many fields, each with its own focus and style, and it is a key part of any public discussion about technology policies.

Because the history of technology is useful in so many different genres, each of the contributors to this volume has used the citation style of their particular academic field, whether it is the footnote-heavy historical approach, the sleek (author, date) method of the social scientist, or even the casual essay form that is deployed in the Afterword. All of these systems and variations are described in *The Chicago Manual of Style* (Chicago: The University of Chicago Press, 14th edition, 1993).

The articles in this book cover a wide range, chronologically, geographically, and philosophically. But as different as they are, they all utilize information and analysis from the history of technology. Our desire has been to show in the following case studies how permeable the boundaries are between theoretically "pure" disciplines and to demonstrate clearly that the history of technology can and must be an important part of all kinds of scholarship, from philosophical speculations on the future of the human to local history to health care and space exploration policy.

The book proper begins with an introductory plea for "Reclaiming Technology for the Humanities" by Carroll Pursell of the History of Technology and Science Department at Case Western Reserve University in Cleveland, Ohio. Carroll is one of the world's leading scholars in the history of technology. Since it was Carroll who brought most of the included authors together in Cleveland for a National

Endowment of the Humanities Summer Seminar on "Technology and American Culture", it is fitting he set this volume's tone.

The first case study is by Gary Edgerton, the chair of the Communication Department at Old Dominion University. He has written a fascinating account of "Digital Color Imaging and the Colorization Controversy: Culture, Technology, and the Popular as Lightning Rod." In it, he shows that the public brouhaha over colorizing old movies that captured so much media attention in the 1980s was hardly the clear-cut white hats (artists) versus black hats (capitalists such as Ted Turner) controversy that it was portrayed to be. While the debate was framed around issues of artistic purity, it was actually driven by questions of profit, loss, and nostalgia. I read with amusement a small notice in my local paper's "Entertainment" section that reported that Ted Turner (called an "evil genius") had abandoned his colorizing (labeled "dubious") campaign after doing only 120 pictures because of a lack of profitability. Having read Gary's article this news, and how it was framed, came as no surprise.

Helping to avoid surprises is one of the goals of the history of technology and few surprises are worse than nuclear accidents. Hence the importance of Monika Ghattas's account of "The Waste Isolation Pilot Plant: Supertechnology in Limbo", a site in Carlsbad, New Mexico that may someday store large quantities of highly radioactive nuclear waste. Monika teaches history at the Division of Arts and Sciences at the Technical Vocational Institute in Albuquerque, New Mexico, so she has been well placed to follow the ups and downs of this project and even to interview some of its key participants. Her detailed history explores the roles of technological optimism and environmental concern in the technical and political debates over this difficult and controversial project.

"Nursing, Technology, and Gender: A History of Progress or Colonization?" is by Lena Sorensen, a professor of nursing at the University of Massachusetts, Boston. She shows through a history of information technologies in nursing that automation doesn't always mean better health care. Since nursing is a "woman's" profession, automating health care has a very different impact on nurses than it does on doctors. Often it includes the deskilling and disempowerment of nurses and even a general decline in the nonquantifiable aspects of nursing (such as nurturing and caregiving) that make hospitals bearable for most of us patients.

Myra Jones, who is a professor in the Department of English, Reading, and Foreign Language at Manatee Community College in

Preface ix

Bradenton, Florida, has written an article about her experiences teaching writing and critical thinking using the history of technology. Specifically she focuses on the work of the journalist Elbert Hubbard, a key turn of the century (nineteenth to twentieth in this case) popularizer of technology. Her emphasis is on teaching students to be better writers through fostering an understanding of the relationship between technology, business, and popular culture by close textual analysis. Her article is called "Promoting Critical Thinking Through Teaching Technohistory."

Next we have Walter Ryan's "Waterpower on the Sugar River in Newport, New Hampshire: A Historical and Industrial Archaeological Investigation." A professor in the Technical Department of New Hampshire Technical College, Walter traces here the history of a series of water-powered mills that since 1766 (the year the town was founded) have played a central role in the life story of Newport, New Hampshire. His is a perfect example of how local history and technological history are often inexorably intertwined.

David Hochfelder, currently finishing his dissertation under Carroll Pursell, has written "Electrical Communication, Language, and Self." He traces how telegraphy fostered "control through communication"—and not just control of markets separated by time and space through institutions such as the Chicago Commodities Exchange, but self-control of the individual through depersonalized conceptions of communication as a purely mathematical process.

My own article, "Medical Cyborgs: Artificial Organs and the Quest for the Posthuman," is a history of those medical technologies that turn people into human-machine systems, or cybernetic organisms (cyborgs). I then try to draw some conclusions about the psychosocial causes, and impacts, of this ever expanding site for transforming the human body.

Then there is a related case study by Heidi J. Figueroa-Sarriera, a professor of social psychology in the Department of Social Sciences at Río Piedras Campus, University of Puerto Rico. Her article, "Some Body Fantasies in Cyberspace Texts: A View from Its Exclusions," is an interdisciplinary analysis of the underlying philosophical and psychological discourses behind virtual reality technologies (also known as cyberspace).

Next come three articles by scholars who weren't participants at the Cleveland NEH seminar. We invited them to join us so that the full range of the history of technology could be demonstrated, since they use it variously in the cause of literary criticism, public policy formation, and social analysis. The first of these "guests" is Steven

Mentor, who is writing his dissertation on the rhetoric of the Information Highway for the English Department of the University of Washington in Seattle. His article, "Manifest(o) Technologies: Marx, Marinetti, Haraway," looks at the role of technology in promulgating the genre of manifestos, and at how many of these manifestos, specifically the Communist Manifesto of Marx, the Futurist Manifesto of Marinetti, and the Cyborg Manifesto of Haraway, have focused on the role of technology in culture as well.

Then we are honored to have a contribution from the director of the History Division for the National Aeronautics and Space Administration (NASA), Roger D. Launius, who discusses the future prospects of NASA in light of its history in "NASA Retrospect and Prospect: Space Policy in the 1950s and the 1990s." His article includes a crucial discussion of the role of history in making public policy.

The last case study is by Gordon Patterson, the editor of the Open Forum Series of which this book is a part. "Color Matters: Race, History, and Florida's Sara Lee Doll" is about an attempt by civil rights activists to create and mass market a realistic "Negro" doll for children in the United States. This story brings up many crucial issues about the role of technology in our own self-understandings as well as technology's part in shaping American, and other, cultures.

Finally, I have used my prerogative as the editor to reflect briefly on the role of the disciplines in the history of technology. I conclude with a few thoughts on the value of the history of technology in interdiciplinary research specifically, and in contemporary scholarship more generally. I hope "Rethinking Technohistory" provokes some thoughts among its readers of the context of contemporary historical research.

All of these articles demonstrate concretely that technology is an integral part of our lives today, of our communities, of our communicating (watching, writing, and networking), of our health (whether it is safety from environmental threats or the desire for a better hospital experience), even of our very images of body and self. Today we invariably must live lives that are always dynamically interacting with multiple and complex technologies. To understand these processes and their ramifications, any serious student of today needs to know something about the technologies of yesterday, not just the mills of two hundred years ago and the telegraph of last century, but also the dolls of a few decades ago, the space flights of a few years ago, and the computer technologies of last year.

The best way to explore the history of technology as a discipline is

Preface

to join the Society for the History of Technology (SHOT, c/o *SHOT Newsletter,* Department of Social Sciences, 1400 Townsend Drive, Michigan Technological University, Houghton, MI 49931-1295), or at least to go to the library and examine some recent issues of SHOT's quarterly journal, *Technology and Culture.* And you can read some of the many fine books from the field.

Each of the articles has its own references but there is also a list of suggested readings at the end of this book. These readings are some of the best books available about the history of technology, with a special emphasis on the American experience and more recent scholarship. Most of these books are not the dry reading you might expect; technohistory is very exciting stuff. But they are hardly simple reading either. It takes a real effort to sort out exactly how technology has, and is, shaping our lives and our society and so the analysis of this process is inevitably detailed. But it is worth the effort. Just remember, you might not be interested in technology and its history, but your life probably depends upon it. Your future certainly does.

ACKNOWLEDGMENTS

The contributors have their own acknowledgements, an important ritual that emphasizes the social and intellectual context of scholarship, but the editor has special thanks to proffer here on behalf of the whole group.

This book would have been inconceivable if Professor Carroll Pursell of the History of Technology and Science Department at Case Western Reserve University had not organized the 1992 National Endowment for the Humanities Summer Seminar on "Technology and American Culture." We also must sincerely thank the National Endowment for the Humanities and all of the taxpayers who make its seminar program possible.

Gordon Patterson, the Open Forum Series editor, inspired this book with his inquiring letter back in the Summer of 1992 and he has helped the book all along the way through its many stages. Mary Roberts was an excellent editor. The staff of the National Aeronautics and Space Agency History Office were also very helpful during visits there.

And finally, the Oregon State University Center for the Humanities has provided invaluable support during the editing of this book. The staff of the center have been extraordinarily kind. Thank you, Peter Copek, director; Laura Rice, acting director; Wendy Madar, coordinator; and especially the late Patty Paulson, administrator; and thank you to the taxpayers of the great State of Oregon.

Chris Hables Gray

CONTRIBUTORS

Gary Edgerton is professor and chair of the Communication and Theatre Arts Department at Old Dominion University. He has written extensively on aspects of the media and cultural history and criticism in a number of books and journals. He is president-elect of the American Culture Association.

Heidi J. Figueroa-Sarriera is a professor of psychology in the Department of Social Sciences, Río Piedras Campus, University of Puerto Rico. Her doctorate is from the University of Puerto Rico, Recinto de Rio Piedras. She is the editor of *Mas Alla de la Bella Indiferencia: Temas in Psicologia de la Mujer* (forthcoming) and is the coeditor of *The Cyborg Handbook* (Routledge Press, 1996). She has written widely on subjectivity, technology, and postmodernism.

Monika Ghattas teaches history at the Technical Vocational Institute in Albuquerque, New Mexico. Her doctorate in history is from the University of New Mexico. Besides the history of technology, her interests include the history of Switzerland, art patronage in Bavaria, and the tradition of dissent in Germany.

Chris Hables Gray is an associate professor of computer science/cultural studies of science and technology at the University of Great Falls in Great Falls, Montana. He was awarded his doctorate from the History of Consciousness Board of Studies at the University of California, Santa Cruz. He is the author of *Postmodern War* (Guilford Press, 1997) and is the editor of *The Cyborg Handbook* (Routledge Press, 1996). Currently he is writing a multimedia work called *The Cyborg Citizen.*

David Hochfelder is a doctoral candidate in the Program in the History of Technology and Science at Case Western Reserve University, Cleveland, Ohio.

Myra Jones teaches composition and literature at Manatee Community College, Venice, Florida. She is exploring the connection between literature and technology, especially in popular culture. She recently presented at the Popular Culture Association Conference on Quantum Mechanics ideas in the Douglas Adams novel, *Dirk Gently's Holistic Detective Agency.*

Roger D. Launius is chief historian of the National Aeronautics and Space Administration. Dr. Launius is the author of numerous articles on the history of aeronautics and space and is the author or editor of ten other books, including *NASA: A History of the U.S. Civil Space Program.* He is presently researching the history of aviation in the American West.

Steven Mentor is a doctoral candidate in the English Department, University of Washington, Seattle. He is the coeditor of *The Cyborg Handbook* (Routledge Press, 1995) and has also written widely on politics, technoscience, and postmodernity. An active poet, he is currently researching the implications (literary and otherwise) of the Information Highway.

Gordon Patterson is a professor of humanities at the Florida Institute of Technology. He has written extensively on technology and the South and is a series editor for Krieger Publishing Company.

Carroll Pursell is the Adeline Barry Davee Professor and director of the Program in the History of Technology and Science at Case Western Reserve University, Cleveland, Ohio. He is a past president of the Society for the History of Technology, the author of many books, and the recipient of many awards for his scholarship and civic contributions.

Walter Ryan is a professor in the Technical Department at the New Hampshire Technical College in Claremont, New Hampshire. He has published widely on the history of New Hampshire and on vocational education. He is a member of the Society of Manufacturing Engineers, the Society for Industrial Archeology, and the New Hampshire Historical Society.

Lena Sorensen has been a nurse since 1970. She also holds three masters degrees and a doctorate in environmental psychology from the City University of New York. She is currently an assistant professor at the University of Massachusetts/Boston, College of Nursing and is the course and clinical coordinator of community mental health nursing there. Besides the role of technology in nursing, she is particularly interested in lesbian health practices.

INTRODUCTION: RECLAIMING TECHNOLOGY FOR THE HUMANITIES
Carroll Pursell

In 1989 an Op Ed piece in *The New York Times* advocated the notion that American colleges should find "a place in the curriculum for the study of technology."[1] Accompanying the article were illustrations of two facades for the halls of learning, both in the form of classical temples. In one the roof was supported by three pillars: Social Science, Natural Science, and the Arts. One corner of the roof, however, was unsupported by these traditional sectors of the General Education requirements. In the second illustration, the sagging corner of the roof was held up by a fourth column, Technology.

The message was obvious and, to me at least, persuasive: in today's society, all of our students, not just those in engineering curricula, need to know about technology. This is not really a radical idea any more, but it is often thought about, I'm afraid, in terms that are defective in at least two ways.

First, the assumption is that engineering students (or those being trained more practically in the mechanical arts) by virtue of their normal curriculum, know something about that great abstraction, Technology. A few years ago when the first set of astronauts were tragically killed in a fire on board a test capsule, public poll-takers discovered that engineers were the *only* segment of American society surprised by the event. Housewives and students, farmers and business people all realized that sooner or later some people were going to get killed because space travel is dangerous and no machine works perfectly all the time. Only engineers, who knew of the many redundant safety systems and the vast odds against an accident happening, were surprised when it did. I take this as evidence that engineers need as much as anyone else to be taught something about the nature of technology.

The other common error lies in thinking that when students are to be taught about technology, it is the technical faculty alone which should do the teaching, on the grounds that they know what they are talking about. I, on the other hand, am inclined to support the proposition that we in the humanities should teach such courses as well, to both technical and arts students, because we also know what we are talking about when it comes to understanding technology.

My general point is that technology is, from one perspective, a form of human behavior, and that what is at stake in technological decisions is what people want and how they will be affected. Technology is not alien to or destructive of our individual and common humanity, it is the very definition of it. We are, simply, animals that use tools. Thus technology is a definition of our humanity, not something foreign to it.

The very word *technology* gained its modern meaning only well into the nineteenth century, and what we now mean by that term would before have been called the "useful arts." We note that the words *art, artificial, artisan, artist,* and *artifact* are all descended from the Latin word *ars*. It should be no surprise that Leonardo da Vinci was both artist and engineer, that Robert Fulton was a professional artist, and that Samuel F. B. Morse was a professor of art.

It was Aristotle who laid down the essential unity of technological practice. Any object, he wrote, combined form, substance, and intention; these three were interrelated; and one could make certain inferences about any of these from a knowledge of any others. To quote him directly, "every *teche* requires the idea of a result before the material realization of that result can be achieved." It is apparent, therefore, that human purpose is at the very heart of the technological enterprise.

In the Early Modern period, when the designers of the Scientific Revolution fought hard to remove purpose as a factor in the scientific understanding of the natural world, the extension of that ban to technology threatened to deprive us of a powerful way of thinking about that subject. It also extended, quite illegitimately I believe, the cloak of neutrality that was thought to cover science, to cover technology as well.

The late social critic Paul Goodman, however, pointed out in 1969 that "whether or not it draws on new scientific research, technology is a branch of moral philosophy, not of science. It aims at prudent goods for the commonweal and to provide efficient means for these goods." Prudence, he reminded us, "is foresight, caution, utility."[2] It was his idea that technologists should exercise this moral restraint, but of course it would help if they were educated to do so, and their

Introduction 3

willingness and ability to do so in no way detract from our responsibility to do the same.

It will sometimes be said that technology should be left to the experts, but if technology is a form of human behavior aimed at the enhancement of people's lives, then the people themselves (that is, all of us) are the real experts in the directions and pace of its activities. We are all the true and final experts in our own lives—what kind of people we want to be, the kind of society in which we want to live—and we cannot escape the responsibility for shaping them as we choose. And technology is a necessary, inevitable, powerful, and useful part of that shaping.

In more than one powerful jeremiad, Lewis Mumford warned against those habits of modern life that "have now produced the catastrophic possibilities, without making the faintest effort to invent the political and moral instruments imperative for their effective control." He wrote of "that qualitative expansion, based on a detachment from organic norms, human purposes, and historic continuities, and the elimination of all other purposes except those proper to science and technology alone. . . ." Finally, he wrote, "to restore a human balance upset by our pathologically dehumanized technology, we must foster human feeling, feeling as disciplined and as refined, by constant application and correction, as our highest intellectual processes."[3]

Mumford's moral challenge is best directed to us, as humanist educators, for it is we who have in our schools and colleges the responsibility for defining, explaining, and justifying those organic norms, human purposes, and historic continuities. It is we who must explain to a generation at risk in a world gravely threatened, that human purpose is not something that must be forced upon modern technology, but rather something that defines it.

Above all, we must refuse the easy assumption that technology is a black box, the secrets within which we can never discover, or a neutral and inevitable force to which we can accommodate ourselves, but never shape to our own ends. We must I think repudiate that motto of Chicago's Century of Progress world's fair held in 1933: "Science Finds—Industry Applies—Man Conforms." Science finds but it also shapes; industry applies, but far more than science; and rather than people conforming to technology, it is technology which conforms to us.

In an Australian short story the author has one character say of another: "I think like a lot of people he's looking for answers instead of questions. But that's the hard thing, isn't it, finding the questions.

Anyone can come up with an answer. Of some sort. You can always dig out an answer from somewhere."[4] It is precisely the task of the humanities to ask those questions rather than put forth answers. About a machine, one can ask a number of questions: who designed it; how does it work; who owns or controls it; what does it do; what does it mean? Perhaps that last question is the most important, and the least asked. We should ask that question of every tool, every machine, every technology with which we interact.

All of us who teach the young (and increasingly the no longer so young) of this nation must accept our own responsibility to understand technology in its full human meaning, and to help our students realize that it is a part of their lives as well, a part for which they have the same responsibility as for any other of their relationships. The members of the 1992 NEH Summer Seminar and their colleagues, whose papers make up this pioneer volume, have taken this charge seriously.

NOTES

1. *The New York Times,* March 29, 1989.
2. Paul Goodman, "Can Technology Be Humane?," reprinted in Albert H. Teich, ed., *Technology and Man's Future* (2d ed., N.Y., 1977), pp. 210, 211.
3. *New York Review of Books,* 6 (April 28, 1966), 3, 5.
4. Marian Eldridge, *The Woman at the Window* (St. Lucia, 1989), p. 180.

1

DIGITAL COLOR IMAGING AND THE COLORIZATION CONTROVERSY: CULTURE, TECHNOLOGY, AND THE POPULAR AS LIGHTNING ROD

Gary Edgerton

COLORIZATION AS A LIGHTNING ROD

The coming of [mass-mediated] color was not purely a triumph. It could also be received as a threatening novelty by those who found safety in words or in older forms of visual representation.

—Neil Harris (1990, p. 319).

All of the visual mass media—lithography, photography, motion pictures, and television—were first realized in black-and-white before being converted to color. In each of these cases, a ground swell of critical controversy accompanied the changeover to chromatic reproduction on the grounds that color was contaminating accepted aesthetic standards, pandering to debased mass tastes, and was primarily the outgrowth of unbridled greed. Every major innovation in the history of motion picture technology has resulted in similar denunciations, especially during the introductions of sound, widescreen cinematography, 3-D, and an assortment of computer-generated special effects, although the extreme level of public commotion surrounding the unveiling of colorization was indeed unique.

Colorization captured America's imagination to a degree that far exceeded its long-term importance as a technique to convert black-and-white films to color on videotape by using a previously developed computer technology known as color imaging. The roots of the impending hostilities actually reach back to early 1981 when the executive board of the Directors Guild of America (DGA) began actively campaigning for a "creative rights" clause in their ongoing contract negotiations with the Motion Picture Association of America (MPAA), the major film producers and distributors's trade association. This inter-industry squabble eventually surfaced, most prominently between 1986 and 1988, becoming the subject of Senate hearings, print and electronic news coverage and editorializing, the TV talk show circuit, speeches by celebrity opinion-leaders, and even a string of jokes delivered by Johnny Carson during several opening monologues on *The Tonight Show,* chiefly at the expense of Atlanta-based media entrepreneur, R. Edward "Ted" Turner, the much-referred-to "mouth of the south."

The most subtle and striking dimension of colorization, in retrospect, was its capacity to serve as a lightning rod, channeling the culture's time-honored ambivalence over the ever growing fusion of art, technology, and commerce into a prolonged and heated debate concerning the ethics of colorizing old monochrome movies. The various factions in the colorization controversy found little common ground amongst themselves, continually resorting to worn-out and transparent stereotypes, which commonly matched high-minded artists against the pervading Philistinism of Hollywood's media moguls. The rhetorical posturing on both sides was, more often than not, disingenuous.

The opening statement by Elliot Silverstein, then-president of the Directors Guild of America (DGA), before a May 1987 Senate judiciary subcommittee on the legality of colorization, for example, aggressively paraded the DGA's idealistic motivations, while concealing the association's more self-serving and longer-standing ambition of residual participation:

> Our compensation, Senator, is not in coin alone . . . our sensibilities are acutely bruised when we see "our children" publicly tortured and butchered on television . . . colorization represents the mutilation of history, the vandalism of our common past . . . the buck is their only bible . . . Mr. Turner, when asked why he was coloring the classic film, "Casablanca," said he was doing it because he "loved the controversy." We find that statement both irresponsible and outrageous (Hearing before the Subcommittee on Technology and the Law, 1988, pp. 3,7,12).

Ted Turner, for his part, exploited the ensuing row with unreserved relish. In June 1988, he made one of his many deliberately provocative announcements concerning the November 9 premiere of a colorized version of *Casablanca* on his superstation, WTBS, to a group of merchandising executives at a business brunch. *Broad-*

casting reported: "an audible gasp was heard from the breakfast audience" ("Here's Looking at Hue, Kid," 1988, p.46). That July Turner candidly admitted: "[*Casablanca*] is one of a handful of films that really doesn't have to be colorized. I did it because I wanted to. All I'm trying to do is protect my investment" (Carter, 1988, p. D5).

Colorization, therefore, is best understood as a localized dispute over profit sharing within the film and television industries which suddenly erupted into a far more public and intractable struggle between competing ideological approaches to contemporary American culture. This conflict provided plenty of melodramatics whereby several seemingly honorable heroes (i.e., Frank Capra, Jimmy Stewart, John Huston, and Woody Allen) confronted a venal, aspiring villain (i.e., Ted Turner) in a media-saturated spectacle, pitting European-based conceptions of art and morality against America's paramount allegiance to the right of private property and its attendant promise of commercial gain. The colorization controversy, in this sense, functions as a kind of popular text, able to be analyzed from a variety of vantage points. My aim, in turn, is to investigate this imbroglio across four interrelated spheres of interest—technology, business and industry, law and policy, and culture—while exploring how developments in each one of these areas impacts on the other three. I will then conclude with some final assessments of why this whole affair remains culturally meaningful.

COMMERCIALIZING COLOR IMAGING

We're providing a service for the entrepreneur who is trying to get the most out of his film investment. What we're doing is for television.
—C. Wilson Markle, president of Colorization, Inc.
(Birchard, 1985, p. 75)

Color imaging was first created in 1971 by C. Wilson Markle, an engineer working at Image Transform, a film and video laboratory in Los Angeles. This technology was fashioned in response to a request from the National Aeronautics and Space Administration a year earlier to color some of its black-and-white Apollo footage in the hopes of generating additional public support by broadcasting the eventual colorized version on network television, as well as showing it around Capitol Hill during NASA's routine lobbying efforts (Sheldon, 1987, pp. 164–165). Markle's work agreement required that he share the rights to all of his research breakthroughs for a decade with Image Transform, at which point he became the sole proprietor. This ten-

8 TECHNOHISTORY

Figure 1.1 Colorization became a reference point throughout American popular culture. (© 1988 S. Gross.)

Digital Color Imaging and the Colorization Controversy 9

year obligation actually inhibited the commercial development of his color imaging process until 1981 when Markle left Los Angeles for Toronto, starting his own company, Mobile Image Canada, Ltd., with the backing of local "financiers, brothers Norman and Earl Glick ... whose business interests includ[ed] entertainment, natural resources and manufacturing" (Sheridan, 1987, p. 59).

Film historian Anthony Slide indicates that there were, in fact, two aborted attempts during the 1960s at coloring black-and-white movies via computer technology, although the results in each case were "reasonably poor" (1992, p. 122). The first satisfactory technique, realized by Markle, involved transferring original monochrome film to videotape; analyzing the footage scene-by-scene on a high-resolution monitor; assigning the appropriate colors from the computer's palette to the specific elements in the first and last frames of each scene, such as skin, hair, eyes, clothes, props, and environmental details; fine-tuning each of these "key" frames from the computer program's 4,096 possible colors and shades; marrying the black-and-white videotape to the electronically generated colors stored in the computer's memory, proceeding from the first "key" frame in each scene to the last, scene after scene, yielding a colorized videotape as the end product ("Advances in Colorization," 1987, p. D7; Klopfenstein, 1991, pp. 4–6).

Markle's color imaging process is, most significantly, designed for an analog computer, meaning that the colors are literally superimposed onto an existing monochrome image, albeit faded out, causing the inaccurate tinting that became associated in the public mind with the earliest colorized videotapes. This substandard coloring, however, had little initial impact on market success. Mobile Image Canada, Ltd. was renamed Colorization, Inc. in 1983, and allied under the Glick corporate umbrella, International HRS Industries, with Hal Roach Studios, an owner of an extensive black-and-white film and television library. Together these two Glick subsidiaries produced the industry's first full-length colorized videotape, *Topper,* which was released in August 1985, earning an astounding $2 million profit through television syndication and the home video market in less than two years. Their second combined effort, Laurel and Hardy's *Way Out West,* was similarly profitable, selling more colorized cassettes in six months at $39.95 than the original black-and-white videotaped version sold at $9.95 during the preceding ten years (Bennetts, 1986, p. C14).

The apparent windfall potential of color imaging encouraged the establishment of three other companies in the mold of Colorization,

Inc. Color Systems Technology (CST), based in the Los Angeles area, became the first American competitor in 1983, relying on an analog coloring system developed and patented by Ralph Weinger, and differing only slightly from the Markle process by the way that the color scheme is overlaid on the black-and-white image (Birchard, 1985, p. 77). Despite having the same basic tinting problems with its product as Colorization, Inc., CST was able to gain a brief foothold over its older rival by its location on the west coast, making contacts and signing multifilm deals with two major Hollywood studios, 20th Century-Fox and MGM-UA, in January 1985. CST's first colorized release, *Miracle on 34th Street,* "was the highest-rated syndicated film [on television in 1986], winning an audience almost three times greater than that of the black and white version the previous year," and thus earning Fox $600,000 in that one showing "or about as much as all the movie's previous years on TV combined" ("Is Movie Colorization a Moral Issue for Broadcasters?" 1987, p. 20; "Companies to Watch: Color Systems Technology Inc.," 1986, p. 45).

The third colorizer, Tintoretto, Inc., was formed in Toronto in 1986 by former employees of Colorization, Inc., who utilized a color imaging process similar to Markle's, since he had never secured a patent on his system. Both Markle and Weinger's analog processes were slowly being improved over the years by trail and error, leaving Tintoretto with the lowest quality results in the business by far. As a consequence, Tintoretto struggled and remained a marginal force, colorizing only two films during its first two years of operation, "while losing $3 million because of mismanagement" (Lev, 1989, p. 37).

American Film Technologies, the most successful of the four companies in hindsight, took full advantage of being the last entry into the colorization marketplace, basically building on the missteps of its three predecessors. AFT was founded in 1987 with branches in both Wayne, Pennsylvania, and San Diego by former stockbroker and independent television producer, George Jensen, Jr. He decided to start AFT after seeing CST's *Miracle on 34th Street:* "The color wasn't great but I thought, 'Boy, improve the color, and you have the makings of a whole new industry.' I knew colorization could be an incredible market and that I wanted in on it" (Jaffe, 1989, p. 41). Jensen began by analyzing the state of colorization as of 1986, concluding that the technology needed upgrading, and the business itself required diversification if it was to maintain long-term viability.

Jensen hired Barry Sandrew, a Harvard medical school neuroscientist and researcher who was using computers to colorize CAT scans and other medical-related graphics, to next develop a practical and high-caliber technique for motion pictures. Sandrew designed a system which literally digitized black-and-white into color, rather than just covering over an original monochrome image, ushering the technology into a more improved, second-generation where tonal quality was greatly enhanced, the number of available colors and shades increased to almost 17 million, and labor costs were cut by two-thirds (Fischetti, 1987, pp. 53–54; Citron, 1990, p. D11). AFT garnered $31 million in orders from Turner Entertainment, 20th Century-Fox, and Republic Pictures during its first year of operations; and by 1990, the company controlled 80 percent of the colorization market worldwide, including an estimated 95 percent of all American business (Dempsey, 1988, p. 43; Hook, 1990, p. 1; "Film 'Colorizer' Says Firm Was Profitable in Fiscal 4th Quarter," 1992, p. A5). Sandrew, moreover, expanded his digital color imaging process to include "paperless," computerized animation, an area where management planned to eventually derive 75 percent of the firm's income ("Companies to Watch: American Film Technologies," 1990, p. 93; "American Film Gets a New Chief Executive," 1991, p. D30).

Colorized movies and TV programs ended up averaging an "80% higher rating [in television syndication] than comparable ones in black and white" for approximately two years through 1987, while being "significantly stronger" in the home video market as well, especially in the midwest and southern United States (Bierbaum, 1986, p. 22; "Is Movie Colorization a Moral Issue for Broadcasters?" 1987, p. 20). Frank Capra, in fact, first approached Wilson Markle about color converting *It's a Wonderful Life* after learning about the color imaging process and the inventor's Toronto-based company in 1983. Colorization, Inc.'s art director, Brian Holmes, prepared a ten-minute test tape of *It's a Wonderful Life* for Capra, resulting in his "enthusiastic agree[ment] to pay half the $260,000 cost of colorizing the movie and to share any profits," along with giving "preliminary approval to making similar color versions of two of his other black and white films, 'Meet John Doe' and 'Lady for a Day'" (Lindsey, 1985, p. C23; Sheridan, 1987, p. 57).

Capra did indeed sign a contract with Colorization, Inc. in 1984, which only became public knowledge after he began waging his highly visible campaign against the colorizing of *It's a Wonderful Life* in the spring of 1985, at which point he felt obliged to maintain that the pact was "invalid because it was not countersigned by his son, Frank Capra, Jr., president of the family's film production company" (Lindsey, 1985, p. C23). Capra went to the press in the first place because Markle and Holmes discovered that the 1946 copyright to the film had lapsed into the public domain in 1974, and they responded by reducing Capra's financial participation, and refusing outright to allow the director to exercise artistic control over the color conversion of his films. As a consequence, Colorization, Inc. alienated a well-known and respected Hollywood insider whose public relations value to the company far outstripped any short term gain with *It's a Wonderful Life,* especially in light of the subsequent controversy and Frank Capra's prominent role in getting it all started. Capra, for his part, followed a strategy of criticizing color conversion mostly on aesthetic grounds, while also stating that he "wanted to avoid litigation . . . at [his] age," and "instead w[ould] mobiliz[e] Hollywood['s] labor unions to oppose" colorized movies "without performers and technicians being paid for them again" (Lindsey, 1985, p. C23).

All told, the bottom line motivation behind the colorization controversy was chiefly a struggle for profits by virtually all parties involved, although the issue of "creative rights" was certainly a major concern for a few selected DGA spokespersons, such as Jimmy Stewart, John Huston, and Woody Allen, among others. The glaring inconsistency, however, lies in the fact that motion pictures had clearly been a television staple for almost four decades, resulting in a vast array of distortions and alterations in the original movies being telecast, such as frame cropping, losses in definition and picture resolution, the censoring of dialogue, indiscriminate cutting, frequent commercial interruptions, panning-and-scanning, and lexiconning.[1] Why then was there such a high degree of righteous indignation over color conversion? The reason resides in the chronic nature of the relationship between the Directors Guild of America and the Motion Picture Association of America who were feuding over the creative community's desire for increased participation in the ever growing earnings from the cable television and home video markets. The then DGA president Elliot Silverstein later admitted as bargaining reached an impasse in late 1985: "There was a direct bridge between those negotiations and the campaign against colorization. We were looking for a platform" (Klawans, 1990, p. 162).

The key point to remember is that, first and foremost, color conversion is about television, not film, and the source of the controversy resides in business and economic realities above all others. By the mid-1980s, motion picture distribution proceeded in descending order through five successive "windows": movie theaters; home video (cassette sales and rentals); pay and basic cable television; network television; and in syndication to individual TV stations.

Within this framework, colorized videotapes were earmarked exclusively for the four post-theatrical markets. The creative community, including the members of the major production guilds, typically received no residual payments, (beyond theaters) for pre-1960 films and television programs, which included most everything being colorized.[2] The copyright owners, on the other hand, the movie studios and their allies, reaped and shared all of the colorization profits amongst themselves.

The DGA, the Screen Actors Guild (SAG), the Writers Guild of America (WGA), and the American Society of Cinematographers (ASC) made a concerted effort between 1983 and 1985 to negotiate participation to no avail. They then decided to shift tactics in early 1986, attacking colorization on aesthetic grounds as Capra was doing, in the hopes of garnering public support and putting the MPAA on the defensive. This strategy did open up legislative and legal options which the guilds eventually pursued, but the combined strength of the movie studios (MPAA), the National Association of Broadcasters (NAB), the Video Software Dealers of America (VSDA), and the Association of Independent Television Stations (INTV) was insistent and irresistible. From a business standpoint, Hollywood's creative community never had a chance when faced with such a formidable industrial alliance (Voros, 1989, pp. 9, 12).

More than anything else, the economics of colorization is symbolic of the triumph of television over motion pictures in the entertainment marketplace. The colorization controversy surfaced in 1986 when total revenues ($4 billion) from the home video market first outgrossed the total theatrical box-office for movies ($3.8 billion); when videocassette recorder penetration passed 45 percent of U.S. households, continuing to soar above 70 percent by 1990; and when on March 25 Ted Turner purchased MGM's film library of over 3,650 (2,200 MGM, 750 pre-1948 Warner Brothers, and 700 RKO) titles for $1.2 billion and brashly began using colorization as a promotional strategy, planning to eventually color convert 10 percent of his holdings for use on his superstation, WTBS, and cable network, TNT (Jameson, 1989, pp. 30–39; A Report of the Register of Copyrights, 1989, pp. 166–168; "Turner: Color It Committed," 1988, p. 45).

By 1988, the colorized movie market had stabilized to a point where television audiences were consistently averaging 7 to 10 percent higher for color converted programming as opposed to black-and-white, while most "home video renters d[id] not care about old movies, whether they [we]re colorized or uncolorized" (Easton, 1988, p. D12; Ferguson, 1991, p. 8; Tyler, 1988, p. 50; Young, 1987, p. 9).

The colorization novelty was slowly wearing off. The demand for colorized films and TV series was also principally a matter of conventional wisdom rather than widespread consumer demand. A group of industry insiders revealed during their testimonies before a 1987 Congressional subcommittee hearing that TV station and cable network executives had actually created and perpetuated much of the need for color conversion by insisting on colorized product, while virtually ignoring black-and-white fare (Hearing before the Subcommittee on Technology and the Law, 1988, pp. 66–91). These programs buyers were, in fact, following a thirty-year-old tradition that was established with the launching of color television when "advertisers decided that they could not have commercials in color while sponsoring a black and white production" because the ads "looked flamboyant" and the programs "cheap" (Harris, 1990, p. 334).

Overall, American Film Technology, Color Systems Technology, Colorization, Inc., and Tintoretto functioned together through the 1980s as a kind of cottage industry where several of the more established Hollywood firms—Fox, Republic, Warner, MGM/UA (and Turner Entertainment Company after March 1986)—simply preferred to farm out small portions of their film libraries to one of the four colorizers in the hopes of sharing in a quick and easy profit. The fact that none of the major motion picture, television, or cable companies ever entered the color conversion business is a conspicuous indication that the market was basically limited, though surviving.[3]

Business factors, in retrospect, were a fundamental part of creating and fueling the colorization controversy, even though they operated mostly beyond the scope of public view. Having been overmatched by a corporate coalition who controlled both the copyrights and the ancillary markets for color converted films and television programs, Hollywood's creative community now turned to the federal government for legal recourse. Among others, Jimmy Stewart argued forcefully on Capitol Hill that colorization was tantamount to "cultural butchery," adding that the colorized version of *It's a Wonderful Life* "broke [Capra's] heart . . . this is a film that is seen every Christmas in America and no one should see it other than the way Frank wanted it to be seen" (McCarthy, 1986, p. 40; "Mr. Smith (and Friends) Come Back to Washington: Color Them Very Upset," 1988, p. 49). The anti-colorization foray was obviously more complicated than the rhetoric suggested. No side was entirely high-minded, as each shared the same general economic goals and assumptions. The studios just happened to be in a better position to dictate the terms of residual participation than were the labor unions. The filmmakers next decided to enter a very different arena where they would again face another well-seasoned, insular, and powerful elite, the members of the United States Congress.

MR. STEWART GOES TO WASHINGTON

In a given battle art may win or it may lose—on the whole it's been admirably tenacious—but artists have generally been wise enough to refrain from calling for help from a government, the one force that's consistently been able to drive art underground.
—Larry McMurtry in an editorial on colorization (1988, p. C7)

Few performers are as beloved in America as Jimmy Stewart, and he was a popular and esteemed visitor during his well-publicized trips to Capitol Hill in March and June of 1988 to campaign against the colorizing of black-and-white movies. The anti-colorization forces had already suffered a series of major setbacks in the previous two years, provoking Stewart to paraphrase his character's words in *Mr. Smith Goes to Washington* (1939): "The only cause worth fighting for is a lost cause" ("Mr. Smith (and Friends) Come Back to Washington: Color Them Very Upset," p. 50; Molotsky, p. C26). He, however, had no way of knowing that there was little legislative leeway left to halt the color imaging of monochrome films by the time of his lobbying junkets, despite continuing gestures from various politicians to the contrary.

This whole controversy is certainly not the first time that members of Congress have used Hollywood's glamour and star power for publicity, and then refused to risk their prestige in the face of even moderate opposition. The maneuvers to involve Washington began in earnest on June 12, 1986, when the Director's Guild of Great Britain issued a statement condemning the coloring of black-and-white features, thus asking the U.S. government to protect a selected number of "classics"—one-hundred American and seventy-five British films—which, in effect, posited the groundwork for the eventual National Film Preservation Act of 1988 (Slide, 1992, p. 129; Greenstone, 1986, pp. 18–20). This declaration signed by eighteen prominent directors invoked a European legal principle, "droit moral" or the moral rights of the artist, which became a flash point for discussion during the ensuing public policy deliberations in Congress.[4]

The concept of moral rights has its legal grounding in an agreement that was created at the 1886 Berne [Switzerland] Convention for the Protection of Literary and Artistic Works. Both the United States and the United Kingdom attended this conference but significantly neither signed the accord, since property rights have always taken precedence over "artistic copyright" in the closely related traditions of Anglo-American jurisprudence (Cooper, 1991, pp. 466–467). Hollywood's labor unions mounted a public relations offensive throughout the summer of 1986 that culminated in a formal request in September by the Directors Guild of America, asking that the Library of Congress refuse to copyright colorized films.

While the Copyright Office considered the DGA's petition, the guild continued its pressure by organizing a highly charged press conference on November 13, 1986, when the octogenarian John Huston, using supplemental oxygen and confined to a wheelchair, responded indignantly to Turner Entertainment's colorizing of *The Maltese Falcon,* a film which he had directed in 1941:

> Last night I looked for as long as I could bear it at a colorized print of 'The Maltese Falcon' . . . an artistic desecration . . . it would almost seem as though a conspiracy exists to degrade our national character . . . the audience is in deepest peril (Robb, 1986, p. 4; Marin, 1986, p. 18).

Huston's fiery remarks addressed the kinds of hot-button issues which stoked the controversy from the beginning. The anti-colorizers were clearly setting the agenda, placing their opponents on the defensive with charges of commercial debasement, despoiling motion picture art, and corrupting the sensibilities of the mass audience. Subsequent commentaries to this effect ranged from two measured, op-ed columns by Woody Allen which argued for "a film artist's moral rights," to Ginger Rogers's wildly histrionic statements before a 1987 Senate judiciary subcommittee hearing on the legality of colorization: "I'm glad that Busby Berkeley isn't here to see what they are doing to his art" (Allen, June 28, 1987; Allen, August 13, 1987, p. 38; Hearing before the Subcommittee on Technology and the Law, 1988, p. 39).

On May 13, 1987, a day after the highly publicized Senate hearing in which Rogers testified along with fellow witnesses Allen, Milos Foreman, and Sydney Pollack, Rep. Richard Gephardt (D-Mo.) introduced H.R. 2400, known as the "Film Integrity Act of 1987 . . . moving against those who would tamper with our American heritage" ("Gephardt fights for black-white movies," 1987). Gephardt's political ambitions were evident in this stratagem, coming as it did in the midst of the 1988 presidential campaign, and more importantly, only days after the withdrawal of Gary Hart from the race. Gephardt's action afforded him the opportunity to both share the spotlight with the famous on Capitol Hill as well as forge a relationship with the film community which until then was overwhelmingly supportive of Hart ("Gephardt Presents Anti-Colorizing Bill," 1987).

Although Gephardt's interest in H.R. 2400 waned with his presidential fortunes, this proposed legislation did guarantee the moral rights of artists by preventing any changes to a film without the consent of the director and screenwriter, or their heirs. The Motion Picture Association of America mobilized a massive lobbying assault in retaliation which effectively stalled any activity on the bill until the following spring. The marketplace alliance of movie studios, broadcasters, and video software dealers was now extending its considerable influence to Capitol Hill en masse.

DGA forces were next frustrated in their efforts to prohibit the copyrighting of colorized films. The Copyright Office of the Library of Congress issued a surprising decision on June 19, 1987, that stated that any individual who added a minimum of three colors to a black-and-white movie can legally copyright the new version as a separate work ("Library of Congress Copyright Office To Register Colorized Motion Pictures," 1987). Ralph Oman, U.S. register of copyrights, expressed his ambivalence over rendering such a judgment in a widely circulated op-ed piece in which he lamented: "I'm smack in the middle of a very American debate—art versus money." Oman then underscored the sovereignty of private ownership in any litigation involving copyright by explaining that "until Congress changes the law to give directors a moral right, I have to apply the existing copyright law" (Oman, 1987).

A second House subcommittee on courts, civil liberties, and the administration of justice was convened in March 1988 in an attempt to resolve the impasse between DGA and MPAA forces, and their respective congressional advocates over H.R. 2400, eventually prompting representatives Robert J. Mrazek (D-N.Y.) and Sidney R. Yates (D-Ill.) to sponsor a proviso to an Interior Department appropriations bill in May as a compromise. The Mrazek-Yates amendment substantially watered down Gephardt's initial proposal by undercutting most of the moral rights considerations in favor of creating a film registry of "national treasures," which would require the labeling of seventy-five movie "classics," if, in fact, these selections were ever colorized or otherwise "materially altered" in any way (see Appendix 1 at the end of this chapter).

Needless to say, the National Film Preservation Act, which President Reagan signed into law on September 27, 1988, created more the perception than the framework for tangible legislative change on the colorization issue (Harris, 1988; "Reagan Signs Law on Film," 1988; Swisher, 1988). In championing the amendment, Mrazek urged his colleagues to "show Jimmy Stewart and the American people that they care about movies" (Schwartz, 1989, p. 145). Some filmmakers, such as then DGA president Elliot Silverstein, were encouraged that "Congress has acknowledged that film is an art," calling it a "historic moment." Woody Allen instantly understood the implications, however, that "unless this is the first crack in the armor that leads to a law protecting the rights of all artists to prevent changes of any type whatsoever to their work without their consent, then I would say it's meaningless" (Yarrow, 1988, p. C13).

Instead of being the proverbial "first crack," Congress quickly retreated on the moral rights issue even further. The National Film Preservation Act of 1988 provided $750,000 over three years to finance a thirteen-member board from the film industry and academia, administered by the librarian of Congress, James H. Billington, for the purpose of building and preserving a canon of distinguished American films.[5] The commission's selections were predictably second-guessed by critics, especially over the preponderance of classical Hollywood movies in the first grouping. After a year, however, board members also began fighting amongst themselves over rules written by Billington, and later inserted into the *Federal Register* on August 9, 1990, which allowed film owners, distributors, exhibitors, and broadcasters "broad leeway to pan and scan, edit and 'lexicon' [registry] films without being forced to label them 'materially altered'" (Wharton and MacBride, 1990).

The librarian of Congress's directive was the result of extensive staff discussions with House members, both those supporting and opposing the Act of 1988, indicating that Congress would not extend the life of the National Film Preservation Act unless the labeling requirements were dropped. Heeding their advice, Billington wanted "to escape the political thicket of film labeling for the relatively quiet pastures of preservation" (Harwood, 1990, p. 4). The House responded by providing a four-year extension for the commission to continue its charge of choosing twenty-five notable films a year with the sole intent of preservation (Wharton, June 17, 1991). In a conciliatory gesture, House copyright subcommittee chair William Hughes (D-N.J.) promised DGA representatives, who vehemently opposed the House's reversal on labeling, to hold a hearing on the moral rights of filmmakers.

The congressional endgame on moral rights began on July 14,

1991, when Rep. Mrazek, who had now emerged as the leading advocate for the DGA, introduced H.R. 3501, the Film Disclosure Act of 1991. This measure was designed to amend present "truth-in-labeling" laws by allowing a film's "artistic authors," limited to the director, screenwriter(s), and cinematographer, to state in a label that they objected to any "material alterations" as a guide to consumers (Gamarekian, 1991; Wharton, July 29, 1991). On March 5, 1992, the first of two days of House copyright subcommittee hearings on moral rights was held with director Martin Scorsese as the featured witness. *Broadcasting* reported that "a majority of [subcommittee members] said they were fans of Scorsese's films, but not the legislation he was promoting" ("DGA Pans Altered States of Movies," 1992).

A follow-up caucus occurred on September 22, 1992, fulfilling Rep. Hughes's pledge to the DGA, but H.R. 3051 never made it out of committee. Most members expressed that they were far more concerned with the bill's potential for increasing the burden of regulation on yet another U.S. industry, than in its capacity to protect the moral rights of film artists. In so doing, the House delivered its final assessment for this session that film is a commodity, first and foremost, subject to the dictates of private enterprise above all else, thus backsliding on the moral rights component in the National Film Preservation Act of 1988. This outcome, moreover, placated MPAA lobbyists who had conducted a massive behind-the-scenes effort to have labeling excised from the legislation.

Although well-meaning, the National Film Preservation Act of 1988 (and its extension) is fundamentally a toothless law, totally ineffectual in addressing moral rights, and producing little more than an elaborate apparatus for naming an annual list of "national treasures." Probably the most hypocritical aspect of the whole episode has been the sponsoring of a commission of "cultural overseers," whose very existence rests on the "preposterous claim about the ignorance and gullibility of the mass film audience" (James, 1989, p. 339). More than one million tax dollars have already been spent underwriting the efforts of the librarian of Congress and the national film preservation board in accordance with the act, or approximately one-third of all the money expended by the federal government on motion picture preservation over this time (Report of the Librarian of Congress, 1993, p. 39). The legacy of the National Film Preservation Act is less a matter of directly conserving the actual films as it is perpetuating the tenure of the commission and upholding its film canon of negligible value and influence.

MUCH ADO ABOUT NOTHING

> The only thing we've been criticized for culturally, to my knowledge, is the colorization of movies. I think that's pretty much a dead issue.
> —Ted Turner (Dawson, 1989, pp. 38–39)

Colorization has long since faded from national attention. The controversy itself was an explosive popular site where forces in business and industry, law and policy, and, most significantly, culture contended with each other, imbuing color imaging with a prominence that far outstripped its practical application as a computer and video technology. No one was more responsible for instigating the tumult, or was better positioned to benefit from its abundant media coverage, than Ted Turner. Even a logical ally like C. Wilson Markle, the inventor of color conversion and an aspiring entrepreneur, was furious with Turner's provocations:

> He went shooting off his mouth about how they were going to colorize *The Maltese Falcon, Casablanca,* and *Miracle on 34th Street*—all the cherished classics. Suddenly the public was in an uproar, and the producers in Hollywood took up their cause. If we all stayed away from coloring the classics, the resistance would diminish (Fischetti, 1987, p. 571).

Turner, of course, encouraged the controversy; it was good for business. He even admitted publicly that he had *The Maltese Falcon* color converted "just for controversy's sake" ("Here's Looking at You, Colorized," 1986). Roger Mayer, president of Turner Entertainment Company, explained "*The Maltese Falcon* was chosen deliberately to create excitement for the opening of the coloring campaign" ("Tainted, Tinted Movies," 1986). Much of the marketing was similarly designed to provoke and inflame. DGA representatives, such as John Huston, impulsively obliged: "[colorization is] as great an impertinence as for someone to wash flesh tones on a da Vinci drawing ("Tainted, Tinted Movies," 1986). Turner reacted with characteristic brashness: "If the director of *The Maltese Falcon* didn't want it colorized, he should have wrote that into the contract when he went to work" ("The More Things Change . . ." 1989).

The spectacle of this bickering, in retrospect, effectively captured the larger, more basic conflict, pitting the homegrown forces of capital against European notions of fine art and culture. Ted Turner, in this sense, was cast as the surrogate Philistine for the whole matrix of film owners, colorizers, cable operators, broadcasters, and video software dealers who regularly intervened between producers and consumers in the film marketplace. He personalized the long-standing process whereby motion pictures were transformed solely into commodities by the mediating power of the movie industry. Turner's statements did nothing more than call attention to conditions that were usually hidden from public view. Such impudence and candor, combined with the irreverence of colorizing, were "like painting a mustache on the 'Mona Lisa' " for many (Darnton, 1986).

Clearly the colorization controversy erupted more as a struggle over the meaning and utilization of movies than anything else. Colorizing the "classics" may have been a publicity stunt for Ted Turner, but it was an assault on tradition for most Americans. The Pantone Color Institute, a nonprofit organization that studies the meaning of color in American life, conducted a survey of 1,200 consumers, including an array of movie critics, in early 1987. The results were decisive: "83 percent of the general [population] and 86 percent of the critics were against the process." According to the institute's executive director, Leatrice Eiseman, "it [wa]sn't just that they were against it that was interesting," but "the vehemence and passion involved" (Darnton, 1987).

The colorization controversy probably produced more heat than light during its peak, serving as a straw man for millions of Americans who seized the opportunity to confirm their "good taste." Over time, though, the accumulated discourse on film art provides a rich cultural record, attesting to the widespread and continuing hold of high modernism on the popular American imagination. Director Jeremy Paul Kagan, for example, likened colorization to "somebody put[ting] blue eyes on 'David' and sa[ying], 'Wouldn't Michelangelo love it?' " (Corliss, 1986). He, like most of his DGA colleagues, muddled the fundamental difference between an art work embodied in a single artifact and film and video which were invented with reproducibility in mind.

German critic Walter Benjamin first delineated this distinction in his seminal 1935 essay, "The Work of Art in the Age of Mechanical Reproduction." Here he discussed "a loss of aura" and a "transformation . . . in our very notion of art," while also being critical of the "doctrine of 'pure' art," or "a theology of art," which he viewed as a romantic retreat and a denial of change (Benjamin, 1969, pp. 217–224). The anti-colorizers were likewise retrospective in their thinking. House member Robert Mrazek, for instance, contended that "great art was being desecrated" when arguing for the passage of the National Film Preservation Act during the summer of 1988. Referring to *The Best Years of Our Lives* (1946) Mrazek continued: "I would rather watch the scene where Fredric March returns to his family than a room full of Renoirs" ("Hope Is Slim for Legislation to Ban Film Coloring," 1988).

Apparently black-and-white films were now "classics" akin to other masterpieces of western civilization, providing yet another object lesson of how some "popular art is transformed into esoteric or high art at precisely that time when it is rendered inaccessible to the types of people who appreciated it earlier" (Levine, 1988, p. 234.) *The New York Times* film critic, Vincent Canby, cautioned against such wholesale canonizing by reminding that "most movies are 'only movies.'" A committed anti-colorizer himself, Canby also expressed his reservations about the commonplace affectations that were becoming de rigueur in the mid-1980s among moviemakers, critics, film scholars, and politicians alike:

> What is troubling about the anti-tinting campaign to date is the tone of righteousness that creeps in, which is harmless in itself, but also promotes a sense of clubby elitism that prevents us from seeing the matter clearly (Canby, 1986, p. 19).

The forces against color conversion were far too precious in their statements to achieve much impact when confronted with the commercial power and political acumen of their opponents. Before a Senate subcommittee, for example, director Milos Foreman merely warned that Americans will "leav[e] the civilized world and enter the jungle" when asked to rebut charges that restricting colorization was a violation of the First and Fifth Amendments.[6] As was typical of DGA witnesses, his arguments relied almost entirely on the strategy of name dropping and the assumed authority of high art: "I realize that I am hired and paid by the money people to make a film, but so was Michelangelo whom Medici hired and paid to paint the Sistine Chapel" (Hearing before the Subcommittee on Technology and the Law, 1988, pp. 34–35). As a result, the anti-colorizers were largely ineffectual in the economic and political arenas, passing only the National Film Preservation Act as a small consolation.

The enduring significance of colorization begins and ends with business and technology. Appendix 2 indicates the proportionally high investment that Ted Turner made in color conversion, having provided more than two-thirds of the product from the outset. The most astonishing figure, in hindsight, is that the whole controversy resulted in just 270 colorized videotapes to date. Recent copyright figures in Appendix 2 also confirm Turner's prediction that colorization is basically a "dead issue." The future of digital color imaging, by contrast, is bright and expanding. Primary system's applications now encompass the work of art conservationists, architects, doctors, interior designers, painters, photographers, sculptors, as well as furnishing new aesthetic vistas for filmmakers and video artists, particularly in the area of "paperless" computerized animation (Armstrong, 1991; Asmus, 1987; Brand, 1987, pp. 219–223; Robertson, 1988).

NOTES

1. Lexiconning refers to the speeding up (time compression) or slowing down (time expansion) of a film shown on television in order to fit the movie into an exact timeslot, while usually adding more commercials in the process. These alterations are accomplished by a Lexicon Time Compressor which is available from the Lexicon Corporation of Waltham, Massachusetts, to any TV station owner who wishes to buy one.

2. Residuals are fees paid to creative talent, including directors, scriptwriters, cinematographers, and performers for additional exposures of their films, TV programs, and even commercials on broadcast TV, cable, or home video. Residual provisions started becoming standard in motion picture contracts after 1960.

3. AFT continues today to maintain its 80 percent share of the colorization market, while Color Systems restructured as CST Entertainment Imaging, Inc. in January 1992, and followed AFT's lead by going digital, thus expanding its repertoire into computer animation and high-definition television (HDTV). Colorization, Inc. and Tintoretto are presently inactive as colorizers, restricted by their analog systems to mainly syndicating their inventories.

4. The statement attacking colorization from the Directors Guild of Great Britain was drafted at the behest of American expatriate, Fred Zinnemann, and signed by Lindsay Anderson, Sir Richard Attenborough, John Boorman, Roy Boulting, Stephen Frears, Hugh Hudson, Roland Joffe, Neil Jordan, Stanley Kubrick, David Lean, Alan Parker, Michael Radford, Karel Reisz, Ridley Scott, John Schlesinger, Michael Winner, and Peter Yates, along with Zinnemann.

5. The thirteen-member National Film Preservation Board is composed of the presidents of the Academy of Motion Picture Arts and Sciences, the Directors Guild of America, the Writers Guild of America, the National Society of Film Critics, the Society for Cinema Studies, the American Film Institute, the University Film and

Video Association, the Motion Picture Association of America, the National Association of Broadcasters, the Association of Motion Picture and Television Producers, and the Screen Actors Guild, as well as the chairs of the film departments at the University of California at Los Angeles and New York University. In 1992, the board was expanded to eighteen with the addition of representatives from the National Association of Theatre Owners, the American Society of Cinematographers and the International Photographers Guild, and the U.S. branch of the International Federation of Film Archives, along with two at-large members.

6. The First Amendment protects free speech. The Fifth Amendment prevents the taking of private property without due process.

REFERENCES

Advances in Colorization: Play It Again, Sam—In Color. (1987, April 22). *The New York Times,* p. D7.

Allen, W. (1987, June 28). The Colorization of Films Insults Artists and Society. *The New York Times,* p. E25.

Allen, W. (1987, August 13). True Colors. *New York Review of Books,* p. 38.

American Film Gets a New Chief Executive. (1991, December 12). *The New York Times,* p. D3.

Armstrong, L. (1991, December 2). 'Lights, Camera . . . O.K., Byte-Brain, Do Your Stuff'. *Business Week,* p. 42.

Asmus, J. (1987, March). Digital Image Processing in Art Conservation. *BYTE,* pp. 151–165.

Benjamin. W. (1969). The Work of Art in the Age of Mechanical Reproduction. In H. Arendt (Ed.), *Illuminations* (pp. 217–251. New York: Schocken Books.

Bennetts, L. (1986, August 5). 'Colorizing' Film Classics: A Boon or a Bane? *The New York Times,* pp. A1, C14.

Bierbaum, T. (1986, December 24). Colorized Movies Find Sales Niche In Homevid Mart. *Variety,* pp. 72–73.

Birchard, R. (1985, October). My Hair is Red, My Eyes Are Blue . . . *American Cinematographer,* pp. 75–78.

Brand, S. (1987). *The Media Lab: Inventing the Future at MIT.* New York: Viking.

Canby, V. (1986, November 30). Through a Tinted Glass, Darkly. *The New York Times,* pp. 19, 24.

Carter, B. (1988, July 28). Ted Turner's New Channel to Air 'Quality' Programs. *The Baltimore Sun,* pp. D1, D5.

Citron, A. (1990, June 20). Cartooning That Could Draw a Suitor. *Los Angeles Times,* pp. D1, D11.

Companies to Watch: American Film Technologies. (1990, June 18). *Forbes,* p. 93.

Companies to Watch: Color Systems Technology Inc. (1986, March 3). *Forbes,* p. 45.

Cooper, R. (Fall 1991). Colorization and Moral Rights: Should the United States Adopt Unified Protection for Artists? *Journalism Quarterly,* 68, 465–473.

Corliss, R. (1986, October 20). Raiders of the Lost Art. *Time,* p. 98.

Darnton, N. (1987, April 17). Debate Goes On Over Colorization. *The New York Times,* p. C8.

Dawson, G. (1989, January/February). Ted Turner: Let others tinker with the message, he transforms the medium itself. *American Film,* pp. 36–39, 52.

Dempsey, J. (1988, April 6). Distribs Find Pics the Color of Money: AFT Pacts for Trio's $31-Mil Order. *Variety,* p. 43, 58.

DGA Pans Altered States of Movies. (1992, March 9). *Broadcasting,* p. 14.

Easton, N. (1988, August 4). Colorization Issue May Be Decided by Committee Today. *Los Angeles Times,* p. D12.

Ferguson, D. (1991, November 3). *The History of Motion Picture Colorization.* Paper presented at the annual meeting of the Speech Communication Association, Atlanta, Georgia.

Film 'Colorizer' Says Firm Was Profitable In Fiscal 4th Quarter. (1992, August 28). *The Wall Street Journal,* p. A5.

Fischetti, M. (1987, August). The Silver Screen Blossoms into Color. *IEEE Spectrum,* pp. 570–576.

Gamarekian, B. (1991, July 25). Film Makers Lobby To Protect Their Work. *The New York Times,* p. 22.

Gephardt Fights for Black-White Movies. (1987, May 13). *San Francisco Examiner,* p. A-3.

Gephardt Presents Anti-Coloring Bill. (1987, May 20). *Variety,* pp. 3, 35.

Greenstone, R. (Fall 1986). A Coat of Paint on the Past?: Impediments to Distribution of Colorized Black and White Motion Pictures. *The Entertainment and Sports Lawyer,* 5, 1, pp. 13–24.

Harmetz, A. (1986, November 14). John Huston Protests 'Maltese Falcon' Coloring. *The New York Times,* p. C36.

Harris, N. (1990). *Cultural Excursions: Marketing Appetites and Cultural Tastes in Modern America.* Chicago: University of Chicago Press.

Harris, P. (1988, October 5). Pic Preservation Board Okayed By Prez; Win Against Colorizing. *Variety,* p. 2.

Harwood, J. (1990, November 5). Film Preservation Bd., Librarian of Congress Talk Future Direction. *Variety,* pp. 4, 16.

Hearing before the Subcommittee on Technology and the Law, Committee on the Judiciary, United States Senate, May 12, 1987. (1988). *Legal Issues That Arise When Color Is Added To Black-and-White Movies.* Washington, D.C.: U.S. Government Printing Office.

Here's Looking at Hue, Kid. (1988, June 13). *Broadcasting,* p. 46.

Here's Looking at You, Colorized. (1986, November 3). *Washington Post,* p. A14.

Hock, S. (1990, April 23–29). What Color was Rick's Place? Ask the Doctor at American Film Technologies. *San Diego Business Journal,* 11, 1, p. 40.

Hope Is Slim for Legislation to Ban Film Coloring. (1988, June 19). *The New York Times,* p. C18.

Is Movie Colorization a Moral Issue for Broadcasters? (1987, May 25). *Broadcasting,* p. 20.

Jaffe, C. (1989, November). The Rainbow Maker. *Nation's Business,* pp. 41–42.

James, D. (July/October 1989). On Colorizing Films: A Venture into Applied Aesthetics. *Metaphilosophy,* 20, 3 & 4, pp. 332–340.

Jameson, R. (1989, July/August). Life with TNT. *Film Comment,* pp. 30–39.

Klawans, S. (1990). Colorization: Rose-Tinted Spectacles. In M. Miller (Ed.), *Seeing Through Movies* (pp. 150–185). New York: Pantheon.

Klopfenstein, B. (1991, November 3). *The Technology of Motion Picture Colorization.* Paper presented at the annual meeting of the Speech Communication Association, Atlanta, Georgia.

Lev, M. (1989, November 11). Little Gold in Coloring Old Films. *The New York Times,* pp. 35, 37.

Levine, L. (1988). *Highbrow/Lowbrow: The Emergence of Cultural Hierarchy in America.* Cambridge: Harvard University Press.

Library of Congress Copyright Office To Register Colorized Motion Pictures. (1987, June 29). *Library of Congress Information Bulletin,* pp. 293–294.

Lindsey, R. (1985, May 19). Frank Capra's Films Lead Fresh Lives. *The New York Times,* pp. C1, C23.

Marin, R. (1986, December 29). Film Snobs Assemble a Color Guard. *The Wall Street Journal,* p. 18.

McCarthy, T. (1986, October 8). AFI Squares Off Against Colorization: Calls for Pros to Get Together. *Variety,* pp. 5, 40.

McMurtry, L. (1988, July 10). The Oranging of James Stewart. *Washington Post,* p. C7.

Mr. Smith (and Friends) Come Back to Washington: Color Them Very Upset. (1988, March 21). *Broadcasting,* pp. 49–50.

Molotsky, I. (1988, June 16). James Stewart's New Role: Lobbyist. *The New York Times,* p. C26.

The More Things Change. . . (1989, November). *American Film,* p. 80.

Oman, R. (1987, June 24). Black and White and Red All Over. *The New York Times,* p. A27.

Reagan Signs Law on Film. (1988, September 28). *The New York Times,* p. C19.

Report of the Librarian of Congress. (June 1993). *A Study of the Current State of American Film Preservation,* Volume 1. Washington, D.C.: Library of Congress.

A Report of the Register of Copyrights. (1989, March). *Technological Alterations to Motion Pictures and Other Audivisual Works: Implications for Creators, Copyright Owners, and Consumers.* Washington, D.C.: U.S. Copyright Office.

Robb, D. (1986, November 19). Huston Blasts Colorizing. *Variety,* pp. 4, 30.

Robertson, B. (1988). Hands-On Digital Retouching. In K. Kelly (Ed.), *Signal: Communication Tools for the Information Age.* New York: Harmony Books.

Schwartz, E. (1989). The National Film Preservation Act of 1988: A Copyright Case Study in the Legislative Process. *Journal of the Copyright Society of the U.S.A.,* 36, pp. 138–159.

Sheldon, K. (1987, march). A Film of a Different Color. *BYTE,* 12, pp. 164–165.

Sheridan, M. (1987, February). Black and White in Living Color. *Canadian Business,* pp. 57–62.

Slide, A. (1992). *Nitrate Won't Wait: A History of Film Preservation in the United States.* Jefferson, NC: McFarland.

Swisher K. (1988, June 30). House Takes Steps on Colorization. *Washington Post,* p. B3.

Tainted, Tinted Movies. (1986, November 16) *The New York Times,* p. E24.

Turner: Color It Committed. (1988, October 24). *Broadcasting,* p. 45.

Tyler, R. (1988, August 3). Retailers Say Black & White More Popular Than Colorized Version. *Variety,* p. 50.

Voros, D. (1989, September 1). Anti-colorization Crusade in Uphill Battle. *Variety,* pp. 9, 12.
Wallace, J. (1986, June 6). Giving New Life to Old Movie Classics. *Publishers Weekly,* pp. 32–36.
Wharton, D. (1991, July 29). Top Directors Get Behind Film-Labeling Legislation. *Variety,* p. 5.
Wharton, D. (1991, June 17). 'Moral Rights' on Hold, as Congress Begins Talks on Preservation Bill. *Variety,* p. 3.
Wharton, D. (1990, August 22). New Federal Rules Give Lotsa Slack For Colorization, Other Pic Alterations. *Variety,* pp. 3, 6.
Wharton, D. and McBride, J. (1990, October 22). Second Group of Pix Chosen for National Film Registry. *Variety,* p. 1.
Yarrow, A. (1988, July 11). Action but No Consensus on Film Coloring. *The New York Times,* pp. C13, C16.
Young, B. (1987, June 13). Video Classics: Color Them Profitable. *Billboard,* p. 9.

APPENDIX 1: THE FIRST 100 FILM "NATIONAL TREASURES"

A 13-member National Film Preservation Board, chaired by the librarian of Congress, James H. Billington, was charged by the National Film Preservation Act of 1988 to select 75 "national treasures" through September 1991 (with an extension thereafter to choose 150 more films over the next six years) that are "an enduring part of our national cultural heritage." Only the first 75 selections cannot be "materially altered" (including colorized) without carrying a label that the changes have been made without the permission of its principal creators.

National Film Registry—1989

1. *The Best Years of Our Lives* (1946)
2. *Casablanca* (1942)
3. *Citizen Kane* (1941)
4. *The Crowd* (1928)
5. *Dr. Strangelove* (1964)
6. *The General* (1927)
7. *Gone With the Wind* (1939)
8. *The Grapes of Wrath* (1940)
9. *High Noon* (1952)
10. *Intolerance* (1916)
11. *The Learning Tree* (1969)
12. *The Maltese Falcon* (1941)
13. *Mr. Smith Goes to Washington* (1939)
14. *Modern Times* (1936)
15. *Nanook of the North* (1922)
16. *On the Waterfront* (1954)

17. *The Searchers* (1956)
18. *Singin' in the Rain* (1952)
19. *Snow White* (1937)
20. *Some Like It Hot* (1959)
21. *Star Wars* (1977)
22. *Sunrise* (1927)
23. *Sunset Boulevard* (1950)
24. *Vertigo* (1958)
25. *The Wizard of Oz* (1939)

National Film Registry—1990

1. *All About Eve* (1950)
2. *All Quiet on the Western Front* (1930)
3. *Bringing Up Baby* (1938)
4. *Dodsworth* (1936)
5. *Duck Soup* (1933)
6. *Fantasia* (1940)
7. *The Freshman* (1925)
8. *The Godfather* (1972)
9. *The Great Train Robbery* (1903)
10. *Harlan County, U.S.A.* (1977)
11. *How Green Was My Valley* (1941)
12. *It's a Wonderful Life* (1946)
13. *Killer of Sheep* (1977)
14. *Love Me Tonight* (1932)
15. *Meshes in the Afternoon* (1943)
16. *Ninotchka* (1939)
17. *Primary* (1960)
18. *Raging Bull* (1980)
19. *Rebel Without a Cause* (1955)
20. *Red River* (1948)
21. *The River* (1937)
22. *Sullivan's Travels* (1941)
23. *Top Hat* (1935)
24. *The Treasure of the Sierra Madre* (1948)
25. *A Woman Under the Influence* (1974)

National Film Registry—1991

1. *The Battle of San Pietro* (1945)
2. *The Blood of Jesus* (1941)
3. *Chinatown* (1974)
4. *City Lights* (1931)
5. *David Holzman's Diary* (1968)
6. *Frankenstein* (1931)
7. *Gertie the Dinosaur* (1914)
8. *Gigi* (1958)
9. *Greed* (1924)
10. *High School* (1968)
11. *I Am a Fugitive from a Chain Gang* (1932)
12. *The Italian* (1915)
13. *King Kong* (1933)
14. *Lawrence of Arabia* (1962)
15. *The Magnificent Ambersons* (1942)
16. *My Darling Clementine* (1946)
17. *Out of the Past* (1947)
18. *A Place in the Sun* (1951)
19. *The Poor Little Rich Girl* (1917)
20. *The Prisoner of Zenda* (1937)
21. *Shadow of a Doubt* (1943)
22. *Sherlock, Jr.* (1924)

23. *Tevya* (1939)
24. *Trouble in Paradise* (1932)
25. *2001: A Space Odyssey* (1968)

National Film Registry—1992

1. *Adam's Rib* (1949)
2. *Annie Hall* (1977)
3. *The Bank Dick* (1940)
4. *Big Business* (1929)
5. *The Big Parade* (1925)
6. *The Birth of a Nation* (1915)
7. *Bonnie and Clyde* (1967)
8. *Carmen Jones* (1954)
9. *Castro Street* (1966)
10. *Detour* (1946)
11. *Dog Star Man* (1964)
12. *Double Indemnity* (1944)
13. *Footlight Parade* (1933)
14. *The Gold Rush* (1925)
15. *Letter from an Unknown Woman* (1948)
16. *Morocco* (1930)
17. *Nashville* (1975)
18. *The Night of the Hunter* (1955)
19. *Paths of Glory* (1957)
20. *Psycho* (1960)
21. *Ride the High Country* (1962)
22. *Salesman* (1969)
23. *Salt of the Earth* (1954)
24. *What's Opera, Doc?* (1957)
25. *Within Our Gates* (1920)

National Film Registry—1993

1. *An American in Paris* (1951)
2. *Badlands* (1973)
3. *The Black Pirate* (1926)
4. *Blade Runner* (1982)
5. *Cat People* (1942)
6. *The Cheat* (1915)
7. *Chulas Fronteras* (1976)
8. *Eaux d'Artifice* (1953)
9. *The Godfather, Part II* (1974)
10. *His Girl Friday* (1940)
11. *It Happened One Night* (1934)
12. *Lassie Come Home* (1943)
13. *Magical Maestro* (1952)
14. *March of Time: Inside Nazi Germany* (1938)
15. *A Night at the Opera* (1935)
16. *Nothing But a Man* (1964)
17. *One Flew Over the Cuckoo's Nest* (1975)
18. *Point of Order* (1964)
19. *Shadows* (1959)
20. *Shane* (1953)
21. *Sweet Smell of Success* (1957)
22. *Touch of Evil* (1958)
23. *Where Are My Children?* (1916)
24. *The Wind* (1928)
25. *Yankee Doodle Dandy* (1942)

Source: Computerized data listing, Motion Picture and Television Reading Room, 336 Madison Building, The Library of Congress, Washington, D.C.

Digital Color Imaging and the Colorization Controversy

APPENDIX 2: COPYRIGHT INFORMATION

Copyright Holders for the 270 Colorized Videotapes Registered Through 31 December 1993

Company	Number	Percentage
Turner Entertainment Company	182	67.4%
Color Systems Technology, Inc.	27	10.0%
Hal Roach Studios	20	7.4%
20th Century-Fox Film Corp.	15	5.5%
Nickelodeon	9	3.3%
American Film Technologies, Inc.	4	1.45%
Republic Pictures Corp.	3	1.1%
RKO Pictures, Inc.	3	1.1%
H. B. Leonard Films	2	0.75%
Sid Luft	1	0.4%
MGM Entertainment	1	0.4%
Otto Preminger Films & Hal Roach Studios	1	0.4%
Vestron Video	1	0.4%
Walt Disney Company	1	0.4%
	270	100.00%

Annual Number of Copyrighted Colorized Videotapes Registered Through 31 December 1993

Year	Number	Percentage
1993	2	0.8%
1992	32	11.9%
1991	58	21.5%
1990	49	18.1%
1989	41	15.2%
1988	49	18.1%
1987	26	9.6%
1986	12	4.4%
1985	1	0.4%
	270	100.0%

Annual Number of Copyrighted Colorized Videotapes—Feature Films to Television Programs—Through 31 December 1993

Year	Number	Features	to TV Programs	Percentage
1993	2	2	0	100 to 0%
1992	32	26	6	81.3 to 18.7%
1991	58	43	15	74.1 to 25.9%
1990	49	49	0	100 to 0%
1989	41	41	0	100 to 0%
1988	49	48	1	98 to 2%
1987	26	26	0	100 to 0%
1986	12	12	0	100 to 0%
1985	1	1	0	100 to 0%

Source: Computed from data obtained at the Copyright Office, Motion Picture, Broadcasting and Recorded Sound Division, Madison Building, The Library of Congress, Washington, D.C.

2
THE WASTE ISOLATION PILOT PLANT: SUPERTECHNOLOGY IN LIMBO

Monika Ghattas

In 1988 the Department of Energy completed excavation of the nation's first mined repository for the permanent storage of radioactive wastes in Carlsbad, New Mexico. Since that time the department has been involved in a lengthy dispute with several government agencies, state officials, scientific groups, and a considerable number of public advocacy groups over its plans to test the safety of this facility by bringing nuclear wastes to the Waste Isolation Pilot Plant (WIPP) and performing various experiments there. Points of contention covered a wide range of topics and included political, scientific, and regulatory issues. On 21 October 1993, the Department of Energy (DOE) canceled this controversial underground testing program. Energy Secretary Hazel O'Leary explained that this action was the result of various studies that questioned the need of transporting and handling actual drums of toxic waste to test the safety of the repository. Instead, the department will conduct aboveground laboratory tests at locations where nuclear wastes are available, such as the Los Alamos National Laboratory. Commenting that the previous administration had not been sufficiently responsive to the concerns of the scientific community, the Environmental Protection Agency (EPA), and the general public in this matter, O'Leary insisted that this reversal of policy has a sounder scientific basis that will enable the department to meet EPA regulations for the permanent storage of nuclear wastes. Moreover, laboratory tests will save government money and take less time, so that the Carlsbad facility can be operational two years earlier than previously planned. Given events of the past fifteen years, however, it seems highly unlikely that the completion of these experiments will lead to the opening of the repository. New questions about its safety and technical aspects, about environmental projections and new nuclear waste technologies, plus other related topics will probably continue to delay this project. Controversial since its inception, this mined repository, 2150 feet underground, remains empty of nuclear waste while the DOE tries to solve a seemingly endless list of problems.

Nuclear waste disposal is not the first technology confronted by controversy and opposition. Many technologies, especially the more spectacular ones like nineteenth century railroads and the advent of electricity, were met with considerable skepticism, ridicule, and altercation. This initial criticism and disquietude quickly disappeared once these technologies became fully operational. Yet, radioactive waste management has stimulated an ever increasing number of critical and seemingly insoluble problems that have prevented the application of this technology. It remains to be seen whether government frustrations with this project mark the beginning of an age less responsive to technological innovation and scientific advances and more suspicious of technological solutions.

The permanent disposal of radioactive materials has had many difficulties. No other technology has been so intimately and yet ominously linked to the future. Radioactive wastes contain materials that will remain dangerous for thousands of years. Therefore, the safety of future generations must be considered. Predicting site and environmental conditions for such a long time involves many probability factors and some degree of risk. The risk factor involved in the permanent storage, transportation, and handling of these wastes dominates public perceptions and fears about the project; it has become one of the main obstacles to the implementation of the repository.

Nuclear waste disposal is also linked to a number of other controversial issues that affect the future of this technology. Radioactive wastes are a byproduct of nuclear armament production. (The Waste Isolation Pilot Plant is designated as a repository for military medium and low level nuclear wastes.) Given current economic conditions and a changing world political order, nuclear weapons research and manufacturing will not receive as much monetary or public support as in the past. It is clear that the Waste Isolation Pilot Project is opposed by certain groups trying to force an end to nuclear arms production by blocking nuclear waste disposal. The fact that nuclear energy production has been responsible for major environmental damage at both commercial and military installations has also affected the issue of radioactive waste, especially since the same federal agency is in charge of both military weapons production and nuclear waste disposal.

Major social and political changes in the last twenty years have influenced the reception of the nuclear waste facility in southeastern New Mexico. Suspicious about governmental initiatives and programs, citizens feel they should participate in the planning and execution of projects that will affect their lives and those of their children. The government did not recognize and legitimize these public priorities; instead, it continued to operate in a Cold War mode where decision making was restricted to federal agencies and kept from public view. The current deadlock over the project is in large part a result of this reluctance to communicate and consult. An inadequate public policy has been more detrimental to the New Mexico Waste Isolation Pilot Project than any of the technical and scientific problems that have arisen during the last few years.

This essay will examine some of the difficulties that have developed over the waste repository between the Department of Energy and New Mexico state officials on one hand and between federal officials and a variety of citizens' groups on the other. A review of major points of contention between these groups will indicate that scientific and technological problems with the repository have been minor factors in the disputes. Instead, most of the controversies are rooted in social, cultural, and political changes that are difficult to anticipate and control.

BACKGROUND

The disposal of radioactive waste elicited little discussion or active concern in the twenty years after World War II. This benign neglect was related to the special position nuclear energy held in public perceptions. On one hand, it was a radically new and incredibly powerful technology that had produced the country's military superiority and was protecting Western democracies from totalitarian aggression. Consequently, atomic energy was associated with secrecy, sophisticated research techniques, and total control by a government that enjoyed the support and good faith of the great majority of its citizens. On the other hand, atomic energy was an expression of optimism and progress, well exemplified in the "Atoms for Peace" program launched during the Eisenhower administration. The "peaceful atom" promised a future filled with unimaginable possibilities that would bring material prosperity to everyone and dramatically improve the overall quality of life. It was a culture of "technological optimism" with progress as its most important product. Technological alienation had not been articulated and no one seriously questioned the viability of a new technology or invention.[1]

One particularly sensational example of this nuclear enthusiasm was Project Plowshare, a proposal to use controlled nuclear bombing for huge public works projects, such as canal construction, river diversion schemes, and difficult mining explorations. Physicist Edward Teller described plowshare as "geographic engineering" and detailed its incredible potential in articles he wrote for *Popular Mechanics* and *This Week*. Interestingly, some early testing for this project was done in the Carlsbad area in 1961. Project Gnome, the name of the secret underground bomb detonation in New Mexico, left radiation traces that are now interfering with some preliminary studies on radiation levels in the area.[2]

By the mid-1960s this period of unbounded optimism and faith in technological progress came to an end. The public was still enchanted with the spectacular scientific advances of space explorations, but in other areas of life, nuclear technology became linked to growing environmental concerns and ongoing discussions about the quality of life. Above-ground nuclear testing, disturbing revelations about radiation exposure, and the constant proliferation of nuclear energy contributed to the change in attitudes. A clear indication of the shift in public interest was the widespread disquietude that followed Rachel Carson's 1962 publication of *Silent Spring*.

The next decade began with the passage of the National Environmental Policy Act and the first Earth Day—good illustrations of the change in national perspectives. Ralph Nadar called a moratorium on the construction of nuclear power plants and in 1979 the country had its first major nuclear power station accident at Three Mile Island. In the meantime, the federal government realized that the nation's nuclear energy policies needed revision and initiated a series of administrative and policy reforms that continued into the Reagan era.

Riddled by scandals, mismanagement, and repeated violations of basic health and safety standards, the Atomic Energy Commission (AEC) was dismantled in 1974 and two new agencies took its place— the Nuclear Regulatory Commission and the Energy Research and Development Administration. When the Department of Energy was organized in 1977, it assumed the responsibilities of research and development.[3]

In spite of this reorganization and new image, the government's nuclear programs continued to be controversial and often contradictory. To solve overlapping jurisdictions and define agency responsibilities, the Carter administration appointed an Interagency Review Group (IRG). It was to advise the executive branch on the organization of both military and civilian nuclear energy projects; this included the disposal of nuclear wastes. This group completed its report in 1979 and made a number of proposals that would have a direct effect upon the nuclear waste disposal explorations under way in New Mexico at the time.[4]

These explorations were related to a study by the National Academy of Science on radioactive waste disposal. The academy had concluded that mined mineral deposits, especially salt beds, seemed most suitable for long-term storage of nuclear wastes. The search for appropriate sites was begun in earnest after the AEC was forced to move nuclear wastes from its Rocky Flats nuclear production center in Golden, Colorado, to the Idaho Engineering Laboratory, after a fire broke out at the Colorado plant. Concerned Idaho officials were placated by an AEC promise that the wastes would soon be removed to a repository that was already in the planning stages. In the meantime, the Cold War continued and the accumulation of radioactive wastes increased both from military installations and civilian nuclear power plants.[5]

For a brief period government officials investigated the large salt formation in central Kansas for the disposal of nuclear wastes. This interest was expedited by community officials in Lyons, Kansas, when they invited federal officials to study the suitability of old mines in the region for storage. Mining operations in the area were closing down and there was considerable concern about the economic viability of the community. The AEC did some preliminary surveying and soon announced that Lyons would become the first federal repository for radioactive wastes. This preemptive strike by the government caused an immediate uproar in the regional press and among some local government officials. In addition, scientists at the University of Kansas disputed government siting statistics and wondered what government technicians would do about the many open and abandoned mine shafts in the area. Shortly thereafter the AEC canceled the project.[6]

In the meantime, the potash industry in southeastern new Mexico was experiencing difficult times and company officials of United Potash Industries decided to approach the federal government about nuclear waste storage in Carlsbad area salt mines. Tentative investigations of the area followed and by the late 1970s the DOE decided to ask Congress for additional funding to proceed with a thorough study of this site. Not only were government scientists impressed with the technical possibilities of the terrain, but there were also important political and economic considerations involved. These included the relative isolation of the area, the receptive mood of the community, and New Mexico's generally positive position towards all federal projects, even those dealing with nuclear energy.[7] It looked like the government had found a site for its first experimental repository for transuranic nuclear waste twenty-two miles southeast of Carlsbad, New Mexico. In 1979 Public Law 96-164 officially authorized the project and defined its mission "to provide a research and development facility to demonstrate the safe disposal of radioactive waste resulting from the defense activities and programs of the U.S. exempted from regulation by the Nuclear Regulatory Commission."[8]

Accordingly, the Waste Isolation Pilot Plant (WIPP) in New Mexico was planned as a federal project for the disposal of military nuclear wastes. This transuranic waste would be limited to the low- and mid-level radioactive byproducts of nuclear armament productions and would include laboratory equipment, gloves, clothing, and tools contaminated with plutonium and other long-lived substances that may be radioactive for thousands of years. Some hazardous materials, such as cleaning solvents, would also be deposited in disposal drums.[9]

While siting studies continued in Carlsbad in the late 1970s, Carter's Interagency Review Group (IRG) issued its report on the government's nuclear energy program. First of all, the review group recommended that the proposed Waste Isolation Pilot Plant in New Mexico should be placed under the control of the Nuclear Regulatory Commission (NRC). This would force the project to conform to the required licensing regulations of the commission. Also by removing it from control of the military, WIPP would be subject to more public scrutiny and review. The report also suggested that this repository should be designed to hold both civilian and nuclear wastes. This would save government money and simplify administrative procedures.[10]

Congress objected to the commission's report and insisted that WIPP remain a federal project for military wastes. Funded and supervised by the House Armed Services Committee, the New Mexico repository was to remain in government hands, shielded by the customary umbrella of national security and protected from public scrutiny and civilian regulatory agencies that were mandated to be publicly responsive. The NRC had such a mandate.[11]

Shortly thereafter, the Carter administration went a step further and decreed that operations at WIPP cease for the time being. Funds for further siting and other technical investigations were canceled, because the IRG concluded that the Carlsbad siting had been done in undue haste. Other localities must be studied in depth as possible repositories for nuclear waste before a final decision is made.[12] This conflict in the waning months of the Carter administration brought considerable attention to this project. It led to questions about the role of states in siting nuclear waste storage facilities. It also prompted discussion about public participation and consultation in government projects that affect the physical well-being of citizens and involve long-term risks.

PUBLIC POLICY

One of the principal reasons for WIPP's problematic status and poor statewide reception is the absence of a comprehensive public policy that should have been initiated when preliminary studies and excavations were underway in Carlsbad. Even though the DOE has rectified this problem to some extent in recent years, the lack of appropriate communication and consultation with state officials and state residents from the beginning resulted in considerable misunderstanding, confusion, and suspicion about government procedures and decisions.[13]

The lack of a public policy for nuclear waste disposal is related to the close tie between the disposition of radioactive wastes and the nation's nuclear weapons program. Consequently, radioactive waste disposal was included in the tight security umbrella that shielded all military projects. Consultation and notification of the civilian sector were not expected or considered. When the Carter administration suggested that civilian and military wastes be integrated and placed under the control of the NRC, Congress objected, in part, because the commission was mandated to be publicly responsive. Today it is clear that nuclear waste disposal has been affected by this early association with nuclear armaments. The disposal of nuclear waste is much less controversial in several European countries, where nuclear energy first appeared in commercial and nonmilitary applications. Therefore, it has been much easier to design public policy directives there than it has been in the United States.[14]

In addition, the government has had little experience in formulating public policy for new technologies. There were no guidelines or directives to develop such policies for the siting, building, and planning of the Carlsbad repository. Even the 1982 Nuclear Waste Policy Act failed to include some limited opportunities for public participation and consultation.[15] Shifting circumstances regarding nuclear waste disposal and nuclear energy in general have also complicated the task of devising a prudent course of action. A plethora of new agencies, new regulations, and new procedures has overwhelmed and confused federal officials making it difficult to consider and develop an appropriate public policy. Wendell Weart, manager of the Carlsbad project at Sandia National Laboratory, observed a few years ago that no one at the onset of the waste storage project could have imagined all the new laws and regulations for the repository that have been enacted since then.[16]

Nevertheless, given the fundamental changes that had taken place in American government and society by the late 1970s, federal officials should have realized that nuclear waste disposal required a public policy. It was obvious, for example, that most government projects no longer could rely on presumed public acquiescence. New priorities concerned with the general quality of life had replaced an earlier focus on physical well-being. The EPA and the U.S. Freedom of Information Act illustrate the new public awareness and demand for some public accountability if not involvement. Also, the growing discomfort with nuclear energy in general mushroomed with the disclosure of radioactive leakage and accidents at reactor sites, leading many to wonder if they had become the victims of this technology rather than its beneficiaries. Moreover, nuclear waste disposal was an issue unlike any other that preceded it. This was a technology with unseen and unending consequences plus a host of immeasurable factors that nourished public paranoia. The inevitable risk factor associated with nuclear technology, including waste disposal, changed public perceptions and attitudes. As the controversy over nuclear waste disposal grew, it became apparent that there were substantial differences between experts and the public about what constitutes "acceptable risk."[17]

Recent years have seen a noticeable change in DOE public relations. The immense publicity and controversy surrounding nuclear weapons, radiation, and waste disposal have convinced federal officials that the Carlsbad project requires some form of limited public policy. In addition, various citizen groups in the state have launched an aggressive and highly publicized campaign against the nuclear storage facility that has forced the government to answer some of the charges leveled against the project. At the present time, the Westinghouse Corporation, the civilian contractor for the Carlsbad project since 1985, is taking care of most of the public information and education programs related to the repository. It is distributing a variety of brochures, flyers, and fact sheets about the facility and offers educational programs for schools, a speakers' bureau, and other similar programs. One of its most successful public relations projects has been an underground tour of one of the mined chambers. However, these programs are primarily designed for public education and information. Critics of the project feel that the government must provide for more substantial and meaningful public participation, such as citizen intervention in regulation proceedings or in the administration of the repository. There should also be more detailed information about government decision making—the process whereby certain options are eliminated while other are adopted. Early efforts in this direction would have promoted a more constructive dialogue about the whole project and could have prevented some of the current criticism and misunderstanding.

WIPP: THE DOE VERSUS THE STATE OF NEW MEXICO

Relations between the DOE and the state of New Mexico over WIPP have been precarious. This has been especially noticeable in recent years because state officials have taken a more conspicuous and determined position on developing issues. When WIPP was first proposed, the role and jurisdiction of individual states regarding military projects and installations were limited. National security militated against any discussion about states' rights in such matters. New Mexico also has had what recent scholars on the West have defined as a distinctive relationship with the federal government—a result of the prominent role federal agencies have had in state development.[18] The large number of federal installations in the state, including two DOE national laboratories, Sandia and Los Alamos, constitute a major economic factor and obviously affected the state's initial response to the planned nuclear waste repository. However, in the last fifteen years changing perceptions about nuclear waste disposal, major political altercations at home and abroad, plus a large number of new regulations and licensing agencies, have resulted in a much stronger state role in the decision-making process of nuclear waste storage.

The Carter administration first stipulated that states and local governments should participate in the planning and siting of nuclear waste repositories. The states were to exercise a "consultive and concurrence" role according to the president's IRG. This was never satisfactorily defined, but it was decided that states could raise objections at critical points in the project. The possibility of state veto power over nuclear waste storage projects was brought up in 1979 while preliminary sitings were underway in Carlsbad. The DOE assured the state that it would have such a prerogative and that it would be also protected by the licensing of the NRC. In addition, the government agreed to pay the costs for the Environmental Evaluation Group, a WIPP oversight agency designed to protect state and public interest. In March 1979, the New Mexico legislature established both executive and legislative committees to study the Waste Isolation Pilot Plant; it was specifically pointed out that this was to facilitate the state's role to concur on the project as promised by the federal government. However, when Congress authorized WIPP in late 1979, it decreed that this was a special facility under the direct control of the DOE and thus neither subject to the commission nor to the state of New Mexico. This unexplained reversal of policy set the stage for the first problems between the DOE and state officials.[19]

The first serious misunderstanding between state and federal government over the proposed plant in Carlsbad occurred a few days after the Reagan administration took office. Work on the project had come to a halt because the IRG had questioned the design and selection of the Carlsbad site and project funding had been suspended. But three days after the Reagan inauguration, the DOE resumed its exploratory work in Carlsbad. The state of New Mexico was not informed. A surprised Governor Bruce King, as well as state officials and New Mexico's congressional delegation, wondered why they were not advised of this DOE action. Energy officials' reply that "we don't need anything legally or officially from the state," incensed the press, but state officials tried to remain noncommittal. U.S. Senator Pete Domenici even defended DOE's action by observing that "the state's role in WIPP is one of merely keeping informed." U.S. Rep. Joe Skeen appeared even less concerned with administrative procedures; obviously influenced by other factors, he commented that WIPP means "economic security for Carlsbad for hundreds of years."[20]

State unease with the project continued to grow and was further compounded by two new developments. One of these concerned discussions in the energy department about bringing high-level wastes

The Waste Isolation Pilot Plant   43

to the Carlsbad repository for testing. This was a major change in policy as far as state officials were concerned. The governor was still guarded in his criticism of this new DOE plan, but asked that New Mexico be allowed an independent licensing review and the exercise of its concurrence right. The other matter concerned DOE's 1980 *Final Environmental Impact Statement (FEIS)* for WIPP. Governor King found that many of its findings did not agree with state studies on the socio-economic effects of WIPP or with recent technical and scientific findings by state agencies. Points raised by the governor's office included concerns about public safety, adequate emergency planning, and proper licensing of the Carlsbad repository. These same issues continue to the present time.[21]

Rising criticism of DOE in the spring of 1981 led the state's attorney general, now Senator Jeff Bingaman, to ask the U.S. District Court to halt work on WIPP "to vindicate rights guaranteed to the State of New Mexico." The suit also mentioned alleged DOE violations of the National Environmental Policy Act and other statutes. A public opposition group, Citizens Against Radioactive Dumping (CARD), had opened an aggressive campaign against WIPP in February and Bingaman's office had been inundated with anti-WIPP messages for several days before he decided on this legal step.[22]

Ensuing discussions between federal and state officials resulted in the first of several "stipulated" agreements whereby New Mexico was formally included in the development of the 10,240 acre WIPP site. The DOE consented to reevaluate the construction of the nuclear waste repository whenever new data was available. This included an exploratory program then ongoing in Carlsbad, the Site and Preliminary Design Validation. It is interesting that the state's complaints focused on the obvious speed of DOE procedures and decisions. Henceforth, there would be more careful deliberation on DOE operations and a more critical analysis of its findings.[23]

An "Agreement for Consultation and Cooperation" was included in the stipulated agreement and became Appendix A of the above document. This was revised in 1983 to include over sixty "key" events or "milestones" in the construction and operation of WIPP that must be reviewed by the state before they are begun. This addition also included a number of environmental issues. The agreement was further modified in 1984, 1987, and 1988. Not only was the role of the state expanded each time, but other regulatory agencies, such as the EPA, the Department of Transportation, and the NCR were brought into the licensing process.[24]

In the early 1980s Tony Anaya was elected governor and his

stance toward WIPP was noticeably more skeptical than that of his predecessor. Anaya insisted, for example, that the DOE conduct "meaningful" public hearings on their reports, something that had not been brought up before. The familiar issues of transportation, safety, and liability also came up and the governor threatened the Carlsbad operation if it did not address these statewide concerns.[25]

One of the most vexing issues was the required transfer of public land from the Bureau of Land Management to the DOE. By federal law, nuclear wastes could not be stored or deposited on land administered by the bureau. This planned transaction became much politicized by the end of the decade and would lead to another major confrontation between the state of New Mexico and the DOE. In the spring of 1983 the first public hearings on this land withdrawal were conducted in Albuquerque and in Carlsbad. That these were not satisfactory is obvious from statements made by New Mexico's Land Commissioner Jim Baca, who complained that the state was not sufficiently consulted about the planned transfer of more than 10,000 acres. Especially disturbing were the expected losses in mineral, oil, and gas rights—a concern repeated by successive administrations in their discussions with federal authorities.[26]

By 1990 the public land withdrawal issue had grown into a major controversy between the DOE and the state and also between members of New Mexico's congressional delegation. Rep. Bill Richardson felt that the required congressional land transfer should only be carried out if the federal government committed itself on several related issues. These included specific environmental standards, a substantial monetary indemnity for New Mexico, and a much stronger federal commitment to finance the improvement of WIPP related highways in the state. Rep. Joe Skeen, a Republican who represented the district that included Carlsbad, was more interested in a speedy compromise between the state and federal government, so that the completed facility would become operational as soon as possible. He maintained that some safety features and additional storage regulations could be added after WIPP began permanent storage of radioactive wastes; in the meantime, the DOE should be allowed to proceed with its testing without being subject to the environmental standards of long-term storage.[27]

For years the WIPP land transfer question was locked in congressional committees that could not agree on a bill satisfactory to both houses of Congress. Later substantial differences between Senate and House versions of the land transfer bill delayed action further. The House advocated a strong state role in the repository project and

The Waste Isolation Pilot Plant

stricter environmental regulations. New Mexico's two senators, Jeff Bingaman (D) and Pete Domenici (R), finally devised a compromise bill that passed both houses of Congress in the fall of 1992.[28]

The impasse over the land transfer issue was further aggravated by two new officials who came into office in the late 1980s: James Watkins and Tom Udall. Energy Secretary James Watkins, an engineer and former admiral, was asked by President Reagan to head the agency in 1989. This was a critical period for the DOE, because the whole nuclear waste storage issue had become highly politicized with the 1982 Nuclear Waste Policy Act and its revision in 1987. The act stipulated that the Department of Energy develop a list of suitable sites for the storage and disposal of radioactive nuclear wastes. This led to major controversies between the federal government and several states that had been selected for siting. The deadlock between the state of Nevada and the federal government continues over the designation of Yucca Mountain, Nevada, as the first high-level radioactive waste repository. It is expected that the eventual disposition of WIPP will have a major impact on the Nevada controversy. By the late 1980s, the public also was much more vocal about matters involving nuclear energy than it had been only a few years earlier. The chemical accident in Bhopal, India, the 1986 nuclear reactor explosion at Chernobyl, and the end of the Cold War were some of the more obvious reasons for increasing public awareness and concern.

Given these determinants the new energy secretary felt that the best way to reactivate the nation's nuclear energy program was to get the New Mexico repository into operation. Its close ties to federal authority and continuing exemption from the Nuclear Waste Policy Act would facilitate this plan. Watkins quickly initiated a thorough review of the whole project. This included physical and mechanical components as well as all documentation, regulations, and legal decisions. To expedite EPA approval, Watkins worked out a five to seven year testing phase for WIPP. Necessary data would be collected during that time to apply for the required storage permits from the EPA. Then he asked Interior Secretary Manuel Lujan, a native New Mexican, to issue him a temporary land withdrawal permit. And after consulting with several WIPP overview committees, Watkins announced that the first shipments of radioactive nuclear wastes would arrive in Carlsbad sometime in October 1991. Within a few days of this announcement, the state of New Mexico filed an injunction barring the DOE from proceeding with its plans.[29]

This firm stand was articulated by the state's newly elected Attorney General Tom Udall, a nephew of former Secretary of the Interior Morris Udall. Udall had voiced misgivings about the WIPP project during his election campaign and was determined that the DOE follow all proposed regulations and safety procedures. Some of these regulations had not been finalized and would need careful examination by the public and state before their acceptance. Citing a number of technical problems, especially the recent roof collapse in one of the mined WIPP chambers, Udall insisted that WIPP was technically not ready to receive wastes and was even less capable of retrieving them if this would become necessary. Also the temporary land transfer arranged by the Interior Secretary was illegal; this could only be done by Congress. Until then, no nuclear wastes could be put into public lands.[30]

In response to Udall's suit in Washington's District Court a federal judge barred Watkins from transporting wastes to WIPP until the DOE was given permission by Congress. An amended and much compromised WIPP land transfer bill finally passed both houses of Congress in the fall of 1992 and was then signed by President Bush. However, as the *Albuquerque Journal* noted, the controversy over the repository was leaving the political arena for the time being and entering the "regulatory battleground." Although Congress finally gave the energy department jurisdiction over the WIPP site, it modified its self-regulating status by insisting that the department abide by EPA and state regulations regarding the disposal of radioactive nuclear wastes. Since EPA guidelines were thrown out of court in 1987, the agency was instructed to develop new rules appropriate for DOE's widely disputed underground testing program and also for the permanent storage of nuclear wastes. The DOE's failure to submit required information about its testing program to the EPA at the end of August 1993 was the first indication that the department was reconsidering the rationale for its nuclear waste experiments in the Carlsbad repository. With the subsequent cancellation of these tests, the EPA can now work on regulations for long-term storage at WIPP. The energy department, in turn, can focus on some of its other problems.[31]

One of these issues involves the financial arrangements between the federal government and the state of New Mexico. The 1992 land transfer bill included an annual indemnification sum of twenty million dollars for the next 15 years, while nuclear wastes are being deposited in Carlsbad. Part of this money must be used to improve state highways, bridges, and bypasses that will be on the WIPP route. However, these funds are not forthcoming until actual shipments begin, even though state officials insist that this should be done before nuclear waste trucks travel across the state.[32]

New Mexico has become much more adamant about federal compensation to the state. In the early 1980s many state officials welcomed the project as another secure economic boost to the state, already heavily dependent on federal installations. In fact, the rather guarded stance of state officials vis-à-vis DOE decisions was rooted in the fear that too many questions could be misconstrued by federal authorities and the whole project would be canceled. Aware of Carlsbad's economic concerns, energy officials, in turn, included a description of anticipated economic benefits from the repository in their 1980 *Final Environmental Impact Statement.* (The 1990 amended version also includes such a section.) However, subsequent state studies on the economic impact of this project have checked initial optimism and enthusiasm. The New Mexico Energy Research and Development Program concluded that the repository would be a statewide economic asset only if the federal government assumed all costs for emergency preparedness, highway upgrading, and any potential accidents. Critics of WIPP often point out that just a single highway accident involving nuclear wastes will adversely affect both the tourist and the agricultural industry in the state.[33]

State agencies have also become much more concerned about specific safety issues regarding the project. This may be due, in part, to the various citizens' initiatives that have gained considerable support because of their much publicized questions about public safety and "acceptable" risk. State officials have directed their attention primarily to the transportation phase of the project and have questioned federal officials on safety measures and general precautions. In line with these concerns, the state actively participated in discussions about the amount of nuclear waste necessary for DOE's previously planned testing program. DOE figures on the number of barrels that were to be brought to Carlsbad for testing (half of one percent of full capacity) were disputed by several oversight agencies as well as Sandia National Laboratory and the state of New Mexico. Most recently, there has been some discussion about alternate storage sites. The land transfer bill instructed the DOE to identify an alternate site within one year after waste shipments begin. State officials, however, would like the EPA to select an alternate site before anything is brought to Carlsbad—something that could delay the opening for several years more.[34]

In studying the issues that have developed between the state of

New Mexico and the DOE over the nuclear waste repository, it is evident that the state has assumed a much broader jurisdiction than it originally claimed. In the early years, it was primarily a matter of clarifying and defining states' rights in the siting and operation of military installations. Now New Mexico has become a licenser of the facility and one of its principal overseers. In addition, it is clear that federal indemnification and compensation for the state continue to dominate official consideration of the project, even though environmental and sociopolitical topics are also under discussion.

While the final disposition of WIPP is still not clear, the DOE and the state of New Mexico are cooperating in a joint waste technology program that is offered at three state universities, the Navajo Community College, and the two national laboratories. Funded by the DOE, the Waste-management Education and Research Consortium (WERC) has several hundred students enrolled in its program on waste management and environmental restoration. Upon completion of their training, they will be certified as Hazardous Materials Technicians. Carlsbad has also received an Environmental Monitoring Center that is affiliated with WERC and it received twenty-seven million dollars from the DOE to oversee the Carlsbad operation and to build a worldwide environmental monitoring system. Programs like these will strengthen the support for federal projects in the state. It is unfortunate that they were not in place much earlier; now it is unlikely that opponents will be sufficiently impressed to modify their opposition to WIPP.[35]

WIPP AND PUBLIC OPPOSITION

While DOE relations with New Mexico state government officials have been strained at times, they have been outright confrontational with many state residents. According to polls conducted periodically by the state's largest newspaper, *The Albuquerque Journal,* only one third of state residents support the Carlsbad project. There have been no significant changes in these figures over the last five or six years. Obviously the responses are in part the result of how questions are framed and presented, but it seems improbable that a majority would back the project in the next few years. Given these statistics, it seems unlikely that the DOE would agree to a statewide referendum—suggested by some proponents of the repository, who would like to push the project forward.[36]

Not required to initiate a public policy, the DOE began its Carlsbad repository project like any other federal installation. However, federal officials may have anticipated some public opposition, because there was a special effort to keep this operation on a strictly technical and scientific level that would preclude nonprofessional participation. This strong reliance on technical details and scientific data—facts and figures—still characterizes much of the material released by the DOE for public education and information. Sandia National Laboratory in Albuquerque, New Mexico, was selected as the technical advisor for WIPP and its experts were instructed to provide whatever public information was necessary. In addition, the National Academy of Sciences was asked to review pertinent data. Since both of these groups held considerable professional esteem, the DOE may have hoped to minimize or discourage nontechnical inquiries and interest.[37]

In spite of these moves, controversy over the proposed repository erupted soon after the government announced the project in the mid-1970s. In the early years most of the opposition was small, disorganized, and divided by various agendas. Citizens for Alternatives to Radioactive Dumping (CARD), organized in 1978 and still prominent today, saw in WIPP a way to build a statewide constituency on an ever-growing number of environmental issues. Other groups operated on the basis of a shared ideology, such as Citizens Against Nuclear Threat (CANT)—pacifists opposed to anything related to nuclear energy. Before publicly denouncing the project, they spent two months in reading groups studying about radioactive wastes. Since it was difficult to forge a common platform against WIPP, opposition groups attempted to identify the DOE as an outsider determined to destroy New Mexico's traditional lifestyle and natural harmony. This was not a particularly effective strategy, because the federal government countered these moves by insisting that WIPP was a federal project committed to solving a national emergency in nuclear waste that was tied to the overall defense program. Likewise, charges by critics that project planning was influenced by political and nontechnical considerations were quickly dismissed by federal officials and attributed to the hidden—and possibly subversive—agenda of opponents.[38]

The first significant public protest occurred during the 1981 Labor Day weekend when between 100 and 200 people tried to block access to the Carlsbad plant. This was a symbolic protest against WIPP organized during the summer by the "Coalition for Direct Action at WIPP." Twenty-one protesters and eight media representatives were arrested for trespassing. Although the number of participants was negligible, opponents were happy to get some press coverage. Earlier in the same year CARD tried to force the federal government into public hearings after it learned that the Interior Department had issued the DOE a permit to continue with drilling at the Carlsbad site. CARD insisted that a drilling permit implied mineral leasing for which a public hearing was required. This attempt to force the DOE into a public forum failed because the judge disagreed with CARD's interpretation and dismissed the suit.[39]

For the time being, energy officials were successful in circumscribing their opposition and maintaining their technical and scientific pose. When a new shaft was sunk two miles from the original site in 1982, critics charged that this was proof that the site had problems with salt brine and gas. DOE officials calmly denied this charge, but thanked their critics for keeping the agency on its toes. It was obvious that critics did not understand WIPP technical questions, but were using this incident to publicize their antinuclear stand.[40]

WIPP also had its early supporters, who often took charge of the public relations matters that the federal agency failed to address. They issued press releases, sponsored high-school essay contests on nuclear wastes, and distributed pro-WIPP public information pamphlets. Americans for Rational Energy Alternatives (AREA) was especially active in this. They held workshops and seminars to disseminate information on nuclear energy and radioactive wastes. Another support group, New Mexicans for Jobs and Energy, promoted the repository from a more practical angle.[41]

Since the mid-1980s critics have become much better organized and more determined to stop the Carlsbad operation. Moreover, the controversy that surrounded the 1982 Nuclear Waste Act and its amended version five years later contributed to public awareness of nuclear issues and, consequently, accelerated public interest and demands for participation in the New Mexico project. Although WIPP was exempted from Nuclear Waste Act regulations, the questions that surrounded new siting procedures for civilian nuclear wastes led to a reexamination of the Carlsbad site. When Nevada residents rejected the government's selection of Yucca Mountain as the nation's first high-level nuclear waste repository, New Mexico WIPP critics intensified their critique of the Carlsbad location. In addition, revelations about technical problems, such as brine reservoirs and potentially dangerous gases near repository explorations, fueled the opposition's case.[42]

By far the most prominent and vocal WIPP critic is the Southwest Research and Information Center (SRIC) in Albuquerque—a nonprofit educational corporation. Since its beginning in 1971, the center has developed various programs to protect natural resources and promote environmental justice. The center emphasizes citizen participation and its activities include policy analysis, public information, technical assistance, and public skills development. It has also been suggested that the SRIC is interested in social change. In recent years, the center has focused primarily on WIPP and related issues. In the early years of the project, the DOE sometimes was able to disarm SRIC comments by pointing to its lack of technical expertise. This has been corrected, however, and SRIC has organized a very effective public relations operation against the Carlsbad project. The title of its quarterly publication, *The Workbook,* indicates the professional and scholarly image that the center is fostering. Director Don Hancock has become the official spokesman of most WIPP opposition groups. He customarily prepares a reply to all DOE announcements and decisions about the repository and has represented the anti-WIPP case before congressional hearings as well as state legislative committees and other executive agencies. In the 1980s the SRIC devised a very effective strategy for discrediting the government's operation in Carlsbad. Using DOE statistics and documents, it pointed out inconsistencies, faulty figures, and obvious mistakes of federal authorities in their official publications and scientific findings on the project. In recent years, the center has been trying to create the impression that its campaign against the repository has been successful. According to Hancock, it is only a matter of time before the government will cancel and decommission the repository for the very reasons that the SRIC opposed its construction. Consequently, the center will be moving its focus to the larger issue of national nuclear waste disposal.[43]

Another major opposition organization is the Santa Fe based Concerned Citizens for Nuclear Safety (CCNS). Members periodically publish a brochure titled *Radioactive Rag* and maintain a 24-hour Radioactive Hotline plus a weekly CCNS news update on a number of northern New Mexico radio stations. The organization defines itself as nonpartisan and nonprofit dedicated to increase public awareness and citizen involvement in nuclear safety. A description of membership includes most major professional groups with scientists being the first group listed.[44]

The CCNS is a good example of how the risk factor involved in nuclear waste can be manipulated to alarm the general population. A radioactive hotline, operational around the clock, implies constant and unpredictable threat. It also works to keep the topic of nuclear waste foremost in public consciousness and promotes an atmosphere of suspicion and public distrust about the whole issue of nuclear energy.

Early in the project the DOE dismissed its critics by labeling them as antinuclear, antitechnology, antiprogress, and anti-American. This simplistic strategy, however, came to an end when questions concerning the project originated with scientific groups and in circles that had not taken a definitive position on the project, but merely wanted clarification on some points. Their questions revolved around the fact that WIPP was exempted from civilian regulatory standards and that its activities were not subject to independent technical oversight. In August 1987 the Committee to Make WIPP Safe was organized. It not only included many professional people, but was also headed by the former secretary of the state's Health and Environment Department, who had negotiated agreements between the Department of Energy and the State of New Mexico. The committee described itself as a group of New Mexico citizens concerned about but not necessarily opposed to WIPP. They put together a four-point public policy agenda that included EPA approval of WIPP and a guarantee from the federal government that the project's mission would not be expanded to include commercial high-level waste storage or the testing of any high-level radioactive materials.[45]

A Scientists' Review Panel concerned about various technical problems with the site also increased the stature of WIPP opponents. The panel is especially concerned about salt brine that exists in pressurized pools closed to the planned waste site and could eventually mix with radioactive materials and be carried to the surface. SRIC's *The Workbook* describes the panel as a group of concerned mathematicians, hydrologists, geologists, and other scientists. Others, however, dispute the allegedly neutral position of this panel and point out that it was organized—and continues to be influenced—by Dr. Roger Anderson, a respected University of New Mexico geologist, who was upset when his alarm about brine at the Carlsbad site was discounted by the National Academy of Sciences.[46]

Technical problems with the WIPP site, as well as revelations of contamination at other DOE sites, gave status and credibility to DOE critics. But in order to engage a larger percentage of the public in the protest movement, it was necessary to introduce questions and issues that touched New Mexico residents more directly. Technical problems in the repository as well as environmental discussions about the Carlsbad area did not elicit sufficient statewide interest in the project. So while nuclear waste legislation and regu-

The Waste Isolation Pilot Plant   53

lations were debated in Washington, the focus in New Mexico shifted to WIPP safety standards, especially to the transportation of nuclear wastes. Alerted to the alleged danger of radioactive materials on roads and highways, a much larger percentage of the population became involved in the WIPP controversy. In Santa Fe, CCNS concentrated on raising public awareness about nuclear materials passing through. Santa Fe does not have a highway bypass in the northwest and radioactive wastes shipped from Los Alamos national Laboratory—one of the principal producers—would be transported through the city. In 1988, a coalition of capital city businesses joined forces to oppose WIPP—primarily on this issue.[47]

The transportation procedure remains one of the major controversies. Critics have put together figures on the number of shipments that will travel to Carlsbad and have calculated potential accident rates, fatalities, radiation leaks, and so forth. Incomplete emergency training, insufficient first-aid service, unsafe New Mexico bridges, and other related topics are highlighted to underscore the lack of safety. All of this information is hypothetical and involves a large degree of subjectivity. Computations and statistics that are used by WIPP opponents are obviously not based on experience, but are computer generated. Moreover, the probability factor of a highway accident involving nuclear waste is lost in the projection on size and consequences of such a potential disaster. Accident statistics also rarely reflect the number of years covered in the figures. But there is no doubt that the image of nuclear waste drums travelling along state roads has mobilized a large number of people who now actively oppose the repository on that issue alone.

Both sides in the controversy are engaged in a major public information campaign. DOE officials and the Westinghouse Corporation are distributing a large number of flyers, fact sheets, and brochures. They are making no attempt to answer specific charges of WIPP opponents; instead they are trying to explain every facet of the project. Unfortunately, the information distributed is often too technical and data focused to be understood by the average reader. Energy officials are also at a disadvantage, because they must maintain a factual mode of presentation and cannot directly confront their critics. And the strong educational component in the writing frequently results in an overly pedantic and dull presentation. At this point in the controversy, government attempts to present the repository in a favorable light—the underground tours in Carlsbad, for example—are not likely to affect the strength of the opposition, but merely confirm their point of view.

WIPP critics, on the other hand, rely on a strong emotional appeal in everything they produce. This is readily apparent in the titles and striking colors of their publications. The *Radioactive Rag* of the Concerned Citizens for Nuclear Safety has already been mentioned. Recently the All People's Coalition distributed a brightly colored orange leaflet on nuclear sites in New Mexico—its title, *New Mexico: A National Nuclear Sacrifice Zone*. It identified more than a dozen of these zones and warned readers that millions will be endangered by the trucking of 28,868 radioactive waste loads to WIPP. The People's Emergency Response Committee has titled its flyer *The Radioactive Pipeline*. The organization's logo next to the title features two skeletons rotating an atom. The Carlsbad site is never referred to as a waste repository, but always is a nuclear dump—something much resented by WIPP officials who consider the repository a major engineering feat. Anti-WIPP literature relies on a vigorous style of writing and colorful metaphors to broadcast its message. Facts and figures do not need to be documented, while cartoons, diagrams, and drawings are judiciously incorporated into the writing.[48]

For years the DOE refused to participate in a direct dialogue with its critics. In the last few years, however, there have been a few public hearings—often related to the land transfer issue—that have brought WIPP supporters and opponents together. These sessions have not been productive. Often the government's attempt to achieve greater public acceptability by discussion and deliberation merely provides a forum for those questioning the need of the project. Also each side is invariably armed with its own set of "facts" and tries to destroy the other's credibility by linking the opposition to political motives—not difficult to do on either side. A number of hearings have deteriorated into shouting matches and the DOE found it expedient to cancel some of its public information seminars for those reasons.[49]

Although environmental, health, and safety standards continue to be highlighted in all WIPP discussions, there appears to be a subtle reversal of methodology between the DOE and its opponents. In the early years of the project, DOE officials minimized the status and authority of opponents by labeling them as antinuclear, unscientific, and poorly informed. In other words, federal officials effectively marginalized their critics and discounted their significance. Now WIPP critics describe the waste disposal project as an irrelevant and unimportant sideline in the radioactive waste disposal issue. They argue that it does not address the problem of existent nuclear waste, because most of that material was dumped into the ground or in trenches and is not retrievable at the present time. Instead of concentrating on solving the migration of these radioactive and hazardous wastes, they say, the DOE has started this poorly planned project—a costly diversion from the real problems on hand. Moreover, it is ludicrous that the government has spent 1.5 billion dollars on this facility for which the necessary nuclear wastes are not yet available. (More than three-fourth of WIPP's capacity will be used for waste from future nuclear arms production.) In addition, WIPP is nonfunctional, because natural salt formations are not a suitable repository. The best solution for WIPP is to let it decommission itself by pulling all machinery and supports out of the mined chambers, which would lead to its eventual collapse. This well-articulated position by WIPP opponents often suggests that the eventual cancellation of the project is a foregone conclusion.[50]

THE CARLSBAD COMMUNITY'S PLIGHT

The position of Carlsbad residents in the WIPP controversy has been especially difficult. Economic considerations are obviously a major factor in the thinking of this southeastern New Mexico community and the project received more community support here than in any other part of the state. Substantially dependent upon the disappearing potash industry, residents saw in this federal project a way to secure their economic viability. In the early 1970s community officials traveled to Washington and testified about the area's suitability for locating an underground repository. They also cited available local expertise in mining, underground explorations, and similar skills that had come from working in the potash industry. In spite of DOE's insistence that its selection criterion was entirely dependent upon the salt beds in the region, the town's receptive attitude toward the repository was obviously a large factor in its selection.[51]

During the 1980s there was much optimism in the Carlsbad area. First of all, it looked as if the repository would be operational within a few years and that meant economic prosperity for the whole community. Local skills and expertise led to employment opportunities both with WIPP and with related agencies, contractors and other companies affiliated with the project. In fact, many saw themselves as active participants in the project, because they were familiar with mining the salt dome and understood the terminology and procedures for sinking shafts and measuring the physical properties of the terrain.[52]

In recent years this community self-confidence has slowly eroded. The operation has come to a virtual standstill as far as physical explorations and construction are concerned. There is considerable concern that WIPP may not be operational for some time to come. The possibility that it may be canceled altogether is something no one wants to consider for the time being.

The previously planned underground testing program was much criticized in the community. Complaints covered a wide range of issues and these will not be substantially modified by the recent DOE decision to cancel underground experiments. In fact, Carlsbad will see this not only as a further delay in the project, but also as another sinister attempt to exclude the community from meaningful participation. The number of tests previously scheduled at Sandia Laboratory was a major irritant for Carlsbad officials, who want all work done in their community. The Carlsbad mayor, Bob Forrest, even suggested that Sandia employees working on WIPP should be transferred to the repository site, but the research laboratory denied this request citing high costs and other administrative problems. Moreover, the testing program, whether underground or in laboratories, is so technical that local residents feel they are automatically disqualified—a frequent refrain in public complaints about government technical programs. Many in Carlsbad felt that Sandia Laboratory would prolong the testing as long as possible to continue federal funding for the laboratory's part in the project. The community must have similar objectives to experiments now scheduled for Los Alamos. In both cases, the laboratories are free to operate according to their own rules, making it easy to introduce new problems and hypothetical projects that may delay the opening for some time to come. Supporters of WIPP feel that the testing program should have been awarded to a private company working with a budget and subject to a deadline.[53]

Much criticism also is directed toward the type of tests under discussion. One principle objective is to study the path of horizontal gases that may be produced by the inevitable salt compression of wastes in the repository chambers. Critics feel that this is a typical example of hypothetical situations created by the federal government to prolong the test phase and delay the opening. Carlsbad's current director of development calls it "paralysis by analysis." Much more important is the vertical migration of radiation from the shafts that are sunk into the waste chambers. This would have a direct effect upon Carlsbad residents and their descendants. Likewise, the operational part of the project is not being sufficiently tested—

the procedures and mechanics that affect worker safety on a day-to-day basis.[54]

The one bonus that Carlsbad received from WIPP is the Environmental Monitoring and Research Center that was opened recently and is designed to become the world's largest environmental monitoring system. Current tasks are more modest and center on an air and soil testing program to measure radioactivity in the area. Yet even here there is some bitterness, because many feel that the center should have been a DOE priority and not something that took considerable community pressure before it was approved.[55]

Given the initial goodwill of the community, it is surprising that federal officials did so little to ensure the continuation of this support. Instead of including local leaders in decision-making procedures, WIPP officials preferred to operate in their customary secretive mode. In recent years this has been corrected to some extent, but community resentment about this is still noticeable. For example, Carlsbad officials feel they should have been asked to participate in the licensing agencies that were deliberating appropriate standards for WIPP's test phase. There is a widespread feeling that all decisions are made either in Albuquerque or in Washington and that the priorities of Carlsbad are either irrelevant or unimportant. Thus the community is frustrated and resentful that it is caught in the middle between the aloof decision making of the federal government and the determined opposition to the project of many state residents.

CONCLUSION

It is difficult to predict the eventual disposition of the Carlsbad repository. Political and social conditions are changing rapidly and will continue to make an impact on the topic of nuclear waste. Adverse economic developments or specific scientific advances in the field of nuclear waste disposal may force a quick decision. For the present, it is worthwhile to look at some of the central problems that have affected the status of this project and speculate whether some could have been avoided or, at least, minimized. Was it wise to let the Department of Energy with its history of negligence and mismanagement of nuclear armament facilities be in charge of radioactive waste disposal? It is possible that this technology would have received less hostility if it had not been so closely linked to nuclear weapons production. A separation between these two areas would have also permitted more public participation in the siting and development of waste repositories. On the other hand, the growing bureaucracy of the federal government has its own critics; the rationale for creating a federal agency for radioactive wastes may seem much clearer now that it was twenty years ago.

A federally initiated public policy with substantial public participation should have been a priority of government officials. A multifaceted education and information program linked to opportunities for meaningful participation in repository planning could have built a stronger public advocacy for the project. There is no doubt that nuclear waste is a complex issue—difficult to understand and not readily adaptable to democratic principles in decision making. But its link to social and political changes was clear many years ago and the government should have made a stronger effort to inform and include the public in this issue. Obviously, there is the danger that education and information on such a controversial topic may confuse and frighten the public more than it would enlighten it. Yet considerable fear and suspicion about the project have developed in the interim anyway; this is often manipulated by opponents to the project and is now very difficult to confront with factual information and education.

Today the major WIPP issues continue to be safety, health, and environmental considerations. Since these matters are related to the management of radioactive materials, they center on the question of "acceptable" risk. To determine an appropriate definition (or consensus) of what constitutes acceptable risk is not possible right now. The dialogue necessary for such a decision requires an atmosphere of mutual trust and confidence that are both lacking. It is possible that dangerous technologies like nuclear waste disposal need not only a broad public policy, but also a risk management strategy that must include a thorough discussion of potential dangers to public health, general safety, and the environment and of appropriate options for dealing with them. The costs of reducing risk must be carefully and openly evaluated in terms of other fiscal priorities and an acceptable risk factor must be agreed upon by the majority of those involved and affected by this technological implementation. The elimination of all risk is impossible in most modern technologies, but an open discussion about this factor and genuine efforts to reduce it will help in the restoration of public trust.

At the present there are three options for WIPP: opening the repository in the near future; postponing actual waste storage until the project has solved most outstanding issues and has more support; or canceling it altogether. All of these possibilities are controversial and invite new problems. Opening the repository would require an executive directive to do so. Apart from legal technicalities, such a decision would force the government to deal with various adversaries: courts, citizen groups, nationwide environmental and other public advocacy groups, and possibly other government agencies, such as the EPA. It is questionable whether this project is worth such a serious confrontation between the government and a large number of its citizens. It would also endanger the little support the government has on the project. WIPP could not be used as the role model for other repositories and would set a detrimental precedent in the whole topic of nuclear waste storage. Canceling WIPP would be a gross misappropriation of public funds. Even if this underground repository is not the best way to dispose of nuclear wastes, it should be utilized in some way to alleviate the problem. Perhaps it could be used for low-level radioactive wastes or become a regional facility for commercial low-level radioactive waste disposal as specified by the Waste Policy Act. A delay in the opening of the repository may be the best solution at the present time. However, this also involves many problems. Indefinite delays are costly and may, in the end, result in a cancellation anyway. Also, the repository cannot be maintained in its present state for any length of time, because the salt is constantly moving and shifting. It is also questionable if it is possible to build a larger constituency for the repository in the interval—the main justification for such a delay.

Those opposed to the Carlsbad repository have made some intriguing proposals regarding nuclear waste disposal. These include a presidential "Blue Ribbon" committee and a new "Manhattan" project that proponents urge the federal government to initiate. The assumption obviously is that the government can solve this crisis with the careful allocation of money and expertise toward the development of an improved and much safer radioactive waste technology—a sign that the age of technological optimism has not come to a close.

NOTES

1. For general background on nuclear waste disposal see Samuel S. Epstein, Lester O. Brown and Carl Pope, *Hazardous Waste in America* (San Francisco: Sierra Club Books, 1982); Donald L. Bartlett and James B. Steele, *Forevermore: Nuclear Waste in America* (New York: W. W. Norton & Co., 1985); Charles A. Walker, *Too*

Hot to Handle: Social and Policy Issues in the Management of Radioactive Waste (New Haven: Yale University Press, 1983); and League of Women Voters, *The Nuclear Waste Primer: A Handbook for Citizens* (New York: Nick Lyons Books, 1985).

Good for a general review of the atom are Paul Boyer, *By the Bomb's Early Light: American Thought and Culture at the Dawn of the Atomic Age* (New York: Pantheon, 1985) and Michael L. Smith, "Advertising the Atom," in *Government and Environmental Politics: Essays on Historical Developments Since World War II,* ed. Michael J. Lacy (Washington, D.C.: Wilson Center Press, 1989), pp. 233–62.

"Technological optimism" is reviewed in an amusing article by Michael L. Smith, " 'Planetary Engineering': The Strange Career of Progress in Nuclear America." in *Possible Dreams: Enthusiasm for Technology in America,* ed. John L. Wright (Dearborn, Mich.: Henry Ford Museum & Greenfield Village, 1992), pp. 111–23. Other articles in the above book published for a commemorative exhibit of *Popular Mechanics* magazine also dwell on the topic of the technological enthusiasm. See also Thomas P. Hughes, *American Genesis: A Century of Invention and Technological Enthusiasm.* (New York: Penguin Books, 1989).

2. Smith, "Planetary Engineering," pp. 117–20. Also interview with Dean William Hadley, College of Pharmacy, University of New Mexico, 21 July 1993, and U.S., Department of Energy, *Final Supplement Environmental Impact Statement: Waste Isolation Pilot Plant,* Assistant Secretary of Defense Programs, October 1990, p. 4.4.

3. *Nuclear Waste Primer,* pp. 43–44 and Smith, "Advertising the Atom," p. 256.

4. "Presidential Message and Fact Sheet of February 12, 1980," Appendix A in *The Politics of Nuclear Waste,* ed. E. William Colglazier, Jr. (New York: Pergamon Press, 1982), pp. 220–1.

5. U.S. Department of Energy, *Final Environmental Impact Statement: Waste Isolation Pilot Plant,* Assistant Secretary of Defense Programs, October 1980, p. 2.5 and Ted Greenwood, "Nuclear Waste Management in the U.S.," in *The Politics of Nuclear Waste,* p. 32.

6. Don Hancock, "How Not to Find a Nuclear Waste Site," in *The Workbook* 11, no. 2 (April–June 1986), p. 48; Keith Schneider, "Wasting Away," *The New York Times Magazine,* 30 August 1992, p. 56 and Marvin Resnikoff, "When Does Consultation Become Cooption? When Does Information Become Propaganda? An Environmental Perspective" in *The Politics of Nuclear Waste,* p. 191.

7. Ronnie D. Lipschutz, *Radioactive Waste: Politics, Technology, and Risk*. A Report of the Union of Concerned Scientists (Cambridge, Mass.: Ballinger Publishing Co., 1980), p. 144.

8. Panel II, WIPP Exhibit, National Atomic Museum, Sandia Base, Albuquerque, New Mexico.

9. U.S., Department of Energy, *The Waste Isolation Pilot Plant: 1992 Annual Report,* pp. 1–3.

10. "Presidential Message and Fact Sheet," p. 222; *Politics of Nuclear Waste,* p. xviii.

11. Greenwood, "Nuclear Waste Management," p. 56 and Lipschutz, *Radioactive Waste,* pp. 144–7.

12. "Presidential Message and Fact Sheet," p. 222.

13. For information on public policy, public participation, and nuclear waste management, see Michael E. Kraft, "Evaluating Technology through Public Participation: The Nuclear Waste Disposal Controversy," in *Technology and Politics,* eds. Michael E. Kraft and Norman Vig (Durham, N.C.: Duke University Press, 1988); K. Guild Nichols, *Technology on Trial*. Organization for Economic Co-operation and Development (Paris: 1979).

14. Interview of Cas Robinson, Director of National Association of Regulatory Utility Commissioners' New Nuclear Waste Program in *Nuclear Energy Information,* U.S. Council on Energy Awareness (Washington: 1993), p. 7. An interesting comparison of nuclear waste policy between U.S. and Holland is made by Herber Inhaber, "Hands Up for Toxic Waste," *Nature* 249, 6294 (18 October 1990):1231–2.

15. *Nuclear Waste Primer,* p. 77.

16. Neal Singer, a staff writer for the Public Affairs Office at Sandia National Laboratory, noted Weart's remark in "The Hot Topic of WIPP," in *Museum Book* 3, a collection of WIPP related materials in the National Atomic Museum Library, Sandia Base, Albuquerque, New Mexico, n.d., p. 16.

17. For excellent discussions on risk management and the problems involved in restoring public trust, see Paul Slovic, James H. Flynn and Mark Layman, "Risk Perception, Trust and Nuclear Waste," *Environment* 33, no. 3 (April 1991), pp. 7–9, and Howard Kunreuther and Ruth Patrick, "Managing the Risks of Hazardous Wastes," ibid., pp. 12–5.

18. Richard White, "Thrashing the Trails," in Patricia Nelson Limerick, Clyde A. Milner II and Charles E. Rankin, *Trails: Toward a New Western History* (Lawrence, Kansas: Kansas University Press, 1991), p. 38. See also Patricia Nelson Limerick, *The Legacy of*

Conquest: The Unbroken Past of the American West (New York: W. W. Norton & Co., 1987) and Donald Worster, *Under Western Skies: Nature and History in the American West* (New York: Oxford University Press, 1992).

19. "Presidential Message and Fact Sheet," p. 227; *Albuquerque Journal (AJ)*, 18 March 1979, p. A1 and 17 December 1987, p. D4; and Bartlett, *Forevermore,* p. 150.

20. *AJ,* 24 January 1981, p. A1.

21. Greenwood, "Nuclear Waste Management," p. 32 and Emilio E. Varanini, III., "Consultation and Concurrence: Process or Substance," in *Politics of Nuclear Waste,* pp. 147–9.

22. *AJ,* 15 May 1981, p. A6 and 5 February 1981, p. B1.

23. U.S., Department of Energy, *Waste Isolation Pilot Plant (WIPP): FY 1993 Site Specific Plan* (September 1992), pp. 2.1–2.3.

24. Ibid.

25. *AJ,* 2 April 1983, p. B2 and 2 June 1983, p. B2.

26. *AJ,* 1 May 1983, p. A4; 2 June 1983, p. B2; and 5 April 1987, p.C5.

27. *AJ,* 12 July 1987, p. C6 and 28 August 1987, p. B4.

28. *AJ,* 2 August 1992, p. B5; 4 August 1992, A1; 25 September 1992, p. D3; 9 October 1992, p. A1; 16 October 1992, p. A16; and 1 November 1992, p. A1.

29. *AJ,* October 1991, p. A1 and 5 October 1991, p. A1; Schneider, "Wasting Away," pp. 45, 56 and 58.

30. *AJ,* 22 June 1990, p. B3 and Schneider, "Wasting Away," pp. 56 and 58.

31. *AJ,* 30 October 1992; 5 February, p. C3; 11 March 1993; 31 July 1987; and 26 August 1993.

32. Compensation figures were first calculated in the 1987 amendment to the Federal Waste Policy Act. They range between ten and twenty million dollars. In a comparison between Dutch and American nuclear waste policies, Herbert Inhaber comments that this arbitrary sum set by the federal government accentuates the lack of control states have in this situation. In Holland, communities set a compensation price, while the government provides incentives and guidelines. Inhaber, "Hands Up for Toxic Waste," pp. 1231–2; see *AJ,* 1 November 1992, p. A1 for financial details of land transfer bill and 26 January 1988, p. B3; also *Nuclear Waste Primer,* p. 49.

33. Greenwood, "Nuclear Waste Management," p. 32; Ronald Cummings, H. Stuart Burness and Roger G. Norton, *The Proposed Waste Isolation Pilot Project (WIPP) and Impacts in the State of New*

Mexico: A Socio-Economic Analysis, New Mexico Energy Research and Development Program (Santa Fe, April 1981), p. 1.22 and *AJ,* 11 November 1991, p. A3.

34. R. Monastersky, "First Nuclear Waste Dump Finally Ready," *Science News* 140 (12 October 1991), p. 228; R. Monastersky, "More Questions Plague Nuclear Waste Dump," Ibid. 135 (24 June 1989), p. 389; *AJ,* 12 March 1993, p. D3 and 5 May 1993, p. D3.

35. Interview with Dean William Hadley. For more information on WERC, see *New Mexico Business Journal* (February 1991): 25–32 and (February 1992): 22–5.

36. *AJ,* 13 September 1990, p. D1 and 11 April 1992, p. A1.

37. Gary L. Downey, "Structure and Practive in the Cultural Identities of Scientists: Negotiating Nuclear Wastes in New Mexico," in *Anthropological Quarterly* 61 (January 1988):33–4.

38. Ibid., pp. 28–30; *AJ,* 19 November 1978, p. B1; 19 April 1979, p. A1; and 13 January 1981, p. A6.

39. *AJ,* 4 April 1981, p. A8; 8 September 1981, p. A1; 4 December 1981, p. A1; and *Albuquerque Journal Impact Magazine,* 9 September 1982, p. 9.

40. *AJ,* 4 July 1982, p. E1.

41. Downey, p. 31 and *AJ,* 19 November 1978, p. B1.

42. *AJ,* 24 May 1983, p. B2 and 17 December 1987, p. D4.

43. Downey, "Structure and Practice," p. 34; Telephone conversation with Don Hancock, director of SRIC, 5 August 1993. Will Keener, an amused observer of the WIPP controversy, finds that Hancock has a papal anti-blessing for every DOE announcement, see Will Keener, "A Case Study of Rhetorical Argument Expressed in the WIPP Controversy," in *Museum Book 3,* n.d., p. 14. See also *AJ,* 25 May 1989, p. D4.

44. *Radioactive Rag* IV (Winter/Summer 1992), p. 2.

45. Don Hancock, "Getting Rid of the Nuclear Waste Problem: The WIPP Stalemate," in *The Worldbook* 14, no. 4 (October–December 1989), p. 135 and *AJ,* 17 December 1987, p. D4.

46. Hancock, "Getting Rid of the Nuclear Waste Problem: the WIPP Stalemate, p. 135; Downey, "Structure and Practice," pp. 16 and 35; Department of Energy, Albuquerque Operations Office, *DOE-AL News 12,* no. 6 (18 March 1988); *AJ,* 10 May 1988, p. B3 and 11 May 1988, p. D3.

47. *AJ,* 13 February 1988, p. B3; 18 September 1988, p. C1; 29 March 1990, p. D3; 8 April 1990, p. C1; and Don Hancock, "Facts about the WIPP Transportation System," Southwest Research and Information Center, 18 June 1990.

48. Keener, pp. 15–6 and various pamphlets, flyers, and fact sheets mentioned.

49. *AJ,* 10 March 1989, p. B3; 13 June 1989, p. A6; and Downey, "Structure and Practice," passim.

50. Telephone conversation with Don Hancock, director of SRIC, 5 August 1993 and flyers printed by All People's Coalition.

51. Mayor Walter Gerrells testifying before Congress in 1983 claimed that 80–85 percent of the people were behind the project. *Nuclear Info,* (Atomic Industrial Forum, Inc.: Bethesda, Maryland, October, 1983), p. 2. Current mayor, Bob Forrest, increased this figure to 90–95%. Jack Hartsfield, "Carlsbad: Feeling Betrayed," *New Mexico Business Journal* (June 1992), p. 70.

52. Interview with John Heaton, Carlsbad pharmacist and former director of Carlsbad Department of Development, 22 July 1993. Close to one thousand people in Carlsbad are employed by Westinghouse Corporation, the civilian contractor for WIPP. Hartsfield, "Carlsbad," p. 70. See also *AJ,* 22 May 1988, p. C3.

53. *AJ,* 11 June 1993, p. A1 and interview with John Heaton.

54. Interview with John Heaton; see also Hartsfield, "Carlsbad," p. 65.

55. *AJ,* 28 August 1993, p. D3 and interview with Dean William Hadley.

3
NURSING, TECHNOLOGY, AND GENDER: A HISTORY OF PROGRESS OR COLONIZATION?
Lena Sorensen

Technology and gender have an intimate, if complicated relationship. This is especially true of women and technology. Here I'll explore this dynamic within a major occupation of women—nursing. By looking at how the theories and practices of technology have developed in nursing, particularly information technology, we can begin to get a picture of the power of gender, race, class, and sexuality in forming the reality of women's technological experiences. Who more than nurses symbolize the ideal woman—the nurturer, the surrogate mother when you are ill, the comforter and protector—as well as the sexually available pinup. Literature can provide us with a view of these complexities. Nurses are represented in all genres of literature: mystery, romance, science fiction, and horror. The historian Barbara Melosh (1988) discusses the significance of the popular images of nurses, providing us with useful insights into the reality of nursing. She writes:

> In sympathetic portraits, [nurses] are nurturing mothers or caring allies. Sometimes the nurse is the emblem of the outsider, a pathetic spinster or strange recluse. In other depictions, nurses' authority and expertise set them beyond the bounds of proper female behavior: They are shown as icy martinets and sexual predators, threats to male prerogatives.... her knowledge of the body and her proximity to feared illness and death draw her closer to the worlds of forbidden sexuality, human frailty and evil, and supernatural horror. (p. 128)

These images of nurses are not confined to literature but have in many ways shaped the history of the nursing profession since the mid 1800s, and they are no less relevant to the identity of nurses today. As a predominately women's profession, nursing reflects the role of women in our society. Thus by looking at nursing today we can get some insights into the reality of women's lives in this new age of technology.

NURSING HISTORY AND DEMOGRAPHY

Who are nurses? Often when we speak about nursing, we develop an image of a homogeneous group of women in white, at the bedside, skillfully yet gently taking care of a patient. Today nurses, even within the hospital setting, are much more diverse. Differences of class, race, and gender exist within the various nursing roles. There are three general occupational groups of nursing personnel: (1) registered nurses (Rns), (2) licensed practical/vocational nurses (LPNs/VNs), and (3) unlicensed assistive nursing personnel (referred to by many titles: nursing aides, orderlies, home health aides, nursing technicians). It is also important to realize that within these categories are further divisions. There are educational tiers of registered nurses: the diploma graduate, the associate degree, the baccalaureate degree, and the masters degree.

Registered nurses are predominately white women, with approximately 15 percent being from so-called racial minorities. Licensed practical nurses have approximately 25 percent minority representation. Little data is available for unlicensed nursing assistants since they are trained and defined by individual hospital institutions and not subject to national standards (Department of Health and Human Services 1992).

There is a complex history of race and class discrimination within the history of nursing (Hine 1989; Malveaux & Englander 1986; Reverby 1987). At the turn of the century when nursing was struggling to achieve its recognition as a legitimate occupation within health care, the "proper" identity of the women who were admitted into nursing to be trained and practice was strongly influenced by race and class. Black women who wanted to become nurses were not permitted to be trained in the established and accredited hospital training schools. Although Mary Eliza Mahoney is acknowledge to be the first black professional nurse in this country, having graduated from the New England Hospital for Women and Children in Boston in 1879, this hospital had a policy of admitting only "one Negro and one Jewish student" each year (Hine 1989). It was not until 1948 that the American Nurses Association admitted black nurses.

Class also played an important role in the professionalizing of nursing. The role of women as workers presented a problem: proper women (i.e., white middle-class women) were expected to be in the home not in the workplace. Thus nursing believed that to be recognized as a proper women's profession it too had to restrict its members to the proper groups—white women from middle-class families who had been socialized to be submissive and not from classes that encouraged union organizing (Reverby 1987).

Today these class divisions are evidence in the hierarchical educational roles and between the nurses who work on the "floor"—the bedside nurses (RNs, LPNs)—and the nurse administrators. Not only are their paychecks significantly different but their work responsibilities are widely varied. Staff nurses—i.e., those at the bedside—are mainly responsible for patient care and in the daily interactions of the people who work and visit the unit: family and friends of patients, MDs, other health care professionals, and maintenance and housekeeping staff. Administrators, on the other hand, have responsibility for the larger management system, its finances and smooth functioning; they are ultimately accountable to the hospital management and to its fiscal system. Each hierarchical class of nurse has its own role expectations and peer group members. Although areas overlap, the goals of different groups of nurses are potentially in conflict. As in all organizations, management and worker are sometimes at odds.

Even today the implicit goal in the professionalization of nursing (the increased requirements for licensure) is to differentiate itself from the lower classes within the health care hierarchy and associate instead with the upper class, white men in medicine.

NURSING: A WOMAN'S PROFESSION

The priorities and choices made by nursing as a women's profession reflect its history and its relationship to the male dominated medical profession and to the health care industry as a whole. Gender ideology provides us with a framework for analysis of this history.

In "Gender: A Useful Category of Historical Analysis," Joan Scott (1988 p. 42) defines gender as "an integral connection between two propositions: gender is a constitutive element of social relationships based on perceived differences between the sexes, and gender is a primary way of signifying relationships of power." Scott goes on to emphasize the interrelatedness of these two propositions by stating that any change in one is a change in the other. She also expands on the concept of social relationships by identifying *four* interrelated elements that construct what gender is in these relationships:

> [(1) Gender involves] culturally available symbols that evoke multiple (and often contradictory) representations—Eve and Mary as symbols for woman, for example, in the Western Christian tradition—but also, myths of light and dark, purification and pollution, innocence and corruption. ...
> [(2) that there are] normative concepts that set forth interpretations of the meanings of the symbols. ... The position that emerges as dominant, however, is stated as the only possible one. Subsequent history is [then] written as if these normative positions were the product of social consensus rather than of conflict. ...
> [(3) that] this kind of analysis must include a notion of politics and reference to social institutions and organizations. ...
> [(4) that] gender is subjective identity (Scott 1988 p. 43–44).

In nursing, each of these categories of gender ideology finds expression: not only in the contradictory images of the nurse as an angelic, nurturing figure related to the image of the Virgin Mary or on the other hand as "Big Nurse," the character portrayed in *One Flew Over the Cuckoo's Nest,* but also in notions of purity, feminine intuition and caring, and finally in the priorities set by the individual women who choose to enter a so-called "helping" profession. Nursing's status as a woman's field is thus strongly defined by traditional notions of *femininity* and *motherhood.* Almost 97 percent of nurses today are women; historically, one can trace the origin and development of nursing practice to the ideology of women's role in society and work. For this reason, many of the discussions and conflicts that have affected nursing throughout its development have centered around these feminine attributes.

TECHNOLOGY IN NURSING PRACTICE

Information technology has infiltrated every aspect of our lives and has influenced the ways we relate to each other at home, work, school, and at play. Advertisements and professional journals alike proclaim the advancements of this technology and the resultant advancement of life in general. Rarely do we as a public community get the opportunity to question or participate in how this technology is shaped into our lives. Yet as I will discuss, technology does not have a life of its own, evolving out of a void, miraculously producing solutions. Technology is grounded in a social, historical, and political context and thus is shaped within these domains.

Over the past twenty-five years, there has been a significant increase in the utilization of computers in all areas of the hospital, particularly in direct patient care. The use of this new technology in hospitals has been increasing at an extraordinary pace. In 1965 only 8 percent of hospitals utilized any form of computer technology, usually for fiscal reimbursement, supply inventory and ordering, and for patient records. By 1980, almost 70 percent of all U.S. hospitals had hospital information systems up and running (Grier, Ziomek, MacLean & Kim 1985). Today it is easier to count the number of hospitals that do *not* use some form of hospital information systems than to number those that do.

Nursing has been slow to be integrated into this computer network. While 70 percent of hospitals had hospital information systems in place in 1980, only 20 percent of these included any kind of nursing programs (Kjerulff 1988). Today many more hospitals have some form of nursing information system, and many are taking it one step further by emphasizing the need for bedside computers, commonly referred to as point-of-care terminals.

As nursing has responded to the diffusion of computers in health care, there has been much discussion in the nursing literature about the positive potential of this new technology. Computers are often described as the answer to many of nurses' day-to-day frustrations: the solution to many stresses and the medium that would, *finally,* relieve nurses from the drudgery of clerical and routinized work to allow the return to the essence of nursing practice, caring for people. With little empirical data to support these promises, nurses and computer vendors alike nevertheless suggest that computerization will improve the quality of nurses' work and the quality of patient care by facilitating communication and decreasing paper work, thus enabling nurses to more fully realize their "caring" role. The suggested desired result is to enable nurses to have more with patients. Often it is assumed that computers will *directly* effect this change. Because computers manage and organize volumes of information they are widely regarded in nursing as a powerful tool which can ultimately bring about only positive changes in all aspects of work.

The reality of nurses' work experiences today is a glaring contradiction to this promise of relief, greater autonomy and higher status. Were nurses the unknowing victims in the evolution of technology or active participants in the way these systems were adapted to their work as nurses? Has this technology brought real progress or furthered nursing's colonization in the health care industry? I do not want to pose the question "Is the computer *good* or *bad* for nursing?" But rather, I intend to analyze the choices of system models and goals which have accompanied the introduction of computers into nursing practice—the shaping of nursing technology. While many workplace managers and nursing administrators in particular have almost intuitively embraced a scientific model in which standardization and quantification are the top priorities, it is necessary now to question these choices in light of newly available evidence of workplace experiences. It is not the computer per se but the way in which it is used that is at issue here.

Although this chapter specifically discusses information computer systems in nursing, the concept of technology must be defined beyond the material object of the computer terminal. Technology is not merely the physical objects that are designed and used but rather encompasses two additional concepts: (1) a form of knowledge: the 'know how' it takes to design, construct, use, and repair these material objects; and (2) the human activities and practices that make the objects necessary: an object is nothing without this human activity (Wajcman 1991).

Technology is also not neutral. It evolves within a historical and cultural setting: social and political roles are established and affect how this innovation is used and promoted. Without this understanding, we are likely to mythologize its potential as some beneficent force with solutions that are inherently truthful and more virtuous. Technology does not in and of itself determine the change nor does it act unidirectionally on its subjects. Its impact is multidirectional with all parties involved changing, and effecting change.

To understand how information technology has been shaped into the most recent history of nursing, we must examine previous efforts by nursing to adopt new technologies and the evolution of nursing as a health care discipline. As industrialization progressed into the twentieth century, there was a change in how the workplace was managed. Once the machines were developed that streamlined tasks and workplace production, the focus shifted to *management* of the workers who used these machines (Noble 1977). Quantification of work activities and control of workers were the core concepts in Frederick Taylor's development of the scientific management model (Barnes 1983). Originally intended for the factory setting, the scientific management movement—Taylorism—was introduced to hospitals by Lillian and Frank Gilbreth in 1912 in the form of time-motion studies (Reverby 1987). Emphasis was on the identification of the functions and tasks that nurses performed so that they could be measured and adjusted to best increase productivity; it was suggested that this would provide "scientific" evidence for the efficiency and contribution of nurses. Nursing leaders asserted that increased productivity and efficiency would help to get nursing recognized as a profession with appropriate status within the health care industry and in society as a whole. Nursing placed its trust on "scientific" evidence, quantification, as the answer to validation.

The historian Susan Reverby (1987) informs us that the Gilbreths first approached physicians in their attempt to introduce Taylor's scientific management to hospitals. Doctors refused to participate, suggesting that hospital efficiency depended more on nurses than on doctors. The medical profession spent the years that nursing was concentrating on these time-motion studies solidifying its own power within the hospital structure and the health care industry (Reverby 1987). Doctors were not willing to quantify and measure their work, emphasizing instead the unique knowledge of the individual physician. Perhaps they knew what nursing did not, that with quantification and measurement ultimately came increased accountability and supervision from the outside. Nevertheless nursing leaders quickly embraced the scientific management concept. This served two fundamental goals in nursing: first, the desire to become recognized and valued as a profession, restricting nursing to native born, white women from "proper" families (Reverby 1987); and, second, the desire for a system which would establish clear lines of authority over a group that was seen as needing to be disciplined and regulated, *women*. The resulting use of Taylorism and time-motion studies of nursing staff did not empower nurses through improved status or autonomy, but rather served to increase management's control through measures of productivity and efficiency.

These time-motion studies were not limited to the early history of nursing. In the 1950s when the focus of health care changed to look more at the environment's role in human behavior, there was an increased interest in the design of the nursing units. Since the development of the "modern" hospital in the nineteenth century, the typical hospital unit was laid out as a long corridor with patient rooms off each side of the hallway and a nurses station at the ends or junction of the units. The nurses station was and is in reality the only physical space on the hospital unit that is identified as the nurses' space. It is designed and analyzed primarily in its relationship to how it serves the needs of others, how nurse can best serve the needs of patients. In the 1950s the Yale Traffic Index was developed to assess the efficiency of nursing units, using the nurse as the unit of study, translating time-motion studies into the measurement of nurses' steps between the various physical spaces on a unit (Thompson & Goldin 1975). The resultant ranking of the number of steps between units (the fewer the better) became the criteria for the estimation of the efficiency of that unit and implicitly the efficiency and organizational ability of the nurses. Thus a specific design of a unit that incorporated these "scientific criteria" into improving the spatial arrangements should make the nurses on that unit able to function at their best. Yet nurses continue to complain about the stress, burnout, and workload of their working lives.

This history can be understood in the context of Scott's discussion of the ideology of gender. Within labor management, the ideology of logic, the privileging of rationality and measurement, served to represent male status and power. In reality these representations were never used to analyze or describe those actually in power, the physicians. While nurses willingly submitted to increased systematization, doctors insisted on greater autonomy.

These contradictory ideologies are also apparent in the conflict within nursing about *what nursing is:* an art or a science. It is assumed that if nursing can be proven to be a scientific discipline, one with its own theory and measurable activities, then it will achieve equal status and recognition with the predominantly male occupations, specifically medicine. But nursing as a woman's occupation *cannot* ignore, nor will it be *allowed* to ignore, its gendered identity as a profession of intuitive caring, the art of nursing. These conflicting notions, although most actively discussed within nursing itself, serve to maintain power imbalances within the health care industry generally. This contradiction grows out of the assumption that medicine and science are inherently superior to nursing and art. In our society, art is less valued than science, and each is represented as *separate* from the other, not as socially constructed interrelated spheres of activity.

Although the roots of this conflict lie in the development of the nursing profession in the early 1900s, the problem is conceptualized along identical lines today. In an attempt to reformulate the question for this "Information Age," nursing has adopted Naisbeth's concept of "high-tech high-touch." The assumption is that with the increased reliance on technology there will be an almost natural demand for increased human contact, high-touch. Touch, as a significant component of the art of caring, will then become validated in its new relationship to the science of technology.

By asserting that certain aspects of its work are grounded in scientific theory, nursing has sought to acquire one of the critical criteria within this society's norms for professional status. On the other hand, historically, nurses have struggled to maintain the uniqueness of their profession as a woman's profession by asserting that this art of nursing—the caring, humanistic dimension—is a factor that deserves recognition and status as a significant component of nursing work. It is asserted that caring complements the scientific knowledge to make nursing unique. Yet this so-called unique role of nursing as a caring profession has had a more limited effect on elevating the status of nursing in the health care hierarchy today than we might want to believe (Sarason 1985). Coser argued thirty years ago that because, "Our culture is achievement-oriented, so caring for patients is not sufficiently rewarded" (Coser 1963). Yet nursing continues its attempt to get society to value caring today.

INFORMATION TECHNOLOGY IN NURSING TODAY: PROGRESS OR COLONIZATION?

Let us return to the present Information Age and to the original question. How has nursing used this new technology and what impact has it had on nursing practice? Information technology only provides the *means* to do something. How that "something" gets achieved depends on the choices of those who own and develop the computer programs.

Now that technology has moved beyond the ability to measure activity and design physical spaces, and into the manipulation of information, the cure for nursing has refocused on communication and documentation as a way to validate its role. It has been estimated that more than 40 percent of nurses' daily responsibilities is spent in documenting or managing information. If the time that it takes to deal with this information is decreased and if information is more quickly available, then—it is argued—nurses will, finally, be able to provide the kind of care to patients that was originally thought to be the essence of nursing. Langdon Winner calls this "*myth*information":

> the almost religious conviction that a widespread adoption of computers and communications systems along with broad access to electronic information, will automatically produce a better world for humanity. (Cited in Zerzan & Carnes 1991, p. 163)

Winner goes on to say that this "mythinformation" is based on four assumptions: "1) people are bereft of information; 2) information is knowledge; 3) knowledge is power; and 4) increased access to information enhances democracy and equalizes social power" (p. 166). The assumptions that contribute to this "mythinformation" have been present in the nursing literature since the advent of computer information systems into nursing more than twenty years ago. It is frequently asserted that the biggest problem in nursing is the amount of paperwork. Thus it is assumed that the use of computers will relieve many of the stressors that nurses deal with daily.

With the significant changes in health care, nursing has been even more coopted into this "mythinformation" by extending these assumptions further in the definition of the role of nurses. In this capitalist society, value and power are often scaled in *direct* proportion to revenue production. At the same time that computers are more available for nursing to quickly and consistently document the work performed, nursing is also being pressured to develop more creative cost-effective practices. This is in response to the increasing rise in the cost of health care delivery and the resulting fiscal crisis in hospital operation. For example, responding to this pressure and the availability of new information technology, the rationalization in nursing has been that specific functions can now be itemized and documented for reimbursement; no longer should the "cost" of nursing work be part of the general room rate of patients but rather it should be billed and reimbursed as a separate cost just as medical services are.

Many nursing leaders suggest that the invisibility of the monetary value of nursing tasks contributes to the general historical devaluation of nursing in the health care hierarchy. Nursing has approached this challenge in a variety of ways, one of which was to identify measurable standards of nursing care by "pricing its product [and] thereby increasing control over nursing practice" (Johnson 1988, p. 36). Computer information systems were developed that presented nursing care plans using standardized language and limited task choice selections. Much of the early literature introducing nurses to the potential benefits of computerization emphasized relieving the nurse of the excessive time and stress of documentation that has prevented nurses from involvement in the "essence" of nursing, from caring. By standardizing the language and formats of the documentation, time would be saved that could be used in direct contact with patients.

The argument for improved speed in documentation concludes that if nurses have access to accurate information quickly, they will be able to make more accurate decisions, document the results of these decisions faster, and ultimately work more productively. The result is assumed to be greater recognition and status from the health care hierarchy. This expectation is reminiscent of the goals of the scientific management reforms begun in nursing in the early 1900s, now updated and integrated into the Information Age. Just as in 1912, when nursing leaders saw scientific management's time-motion studies as the way to prove the worth of nurses in hospitals, we now see quantification and standardization within information systems rationalized as the way to justify the economic worth and value of nurses.

Originally the computer was introduced as a extraordinary tool that would answer many problems in health care delivery. Some nurses did express their fears and speculate that computers might actually be used to replace them. The argument from many other nurses, however, was that nursing could control the outcomes if they took an active role in the development of information systems (Henderson 1985; Zielstorff 1987) because nurses would identify measurable activities and describe the qualities of nursing work that comprised both the science and art of nursing (Dunlop 1986; Johnson 1988; Larson 1981; Watson 1979). It was assumed that if nurses could control the development of this innovation, then nurses themselves would benefit from automation by the *enhancement* of nursing activities. These tasks were also linked to monetary reimbursement, another factor seen as positive for nurses. Unfortunately nursing has focused this control in the form of measurement and standardization, not enhanced information sharing, dialogue, and group development. These assumptions ignored the power of hospital efforts to increase productivity for increased revenues.

A review of studies that evaluated the outcomes of these computer systems found that the promises were not always achieved. From 1968 to 1989, 119 articles in the major nursing journals dealt specifically with the interaction between computers and nurses in hospital settings. These nursing articles identified only a limited number of benefits of computer use (i.e., decreased time in clerical documentation, increased legibility, increased productivity). Most of these findings had been based on assumptions and not grounded in empirical data; only 24 (20 percent) of the articles cited any actual research to support their conclusions. In the literature that used on-site studies to document outcomes, the results were mixed (Hendrickson & Kovner 1989; Kjerulff 1988; Packer 1986; Staggers 1988), finding that decreased time in documentation did not result in increased time with patients. Sometimes there was actually an increase in workload. The majority of the articles and studies focused on the clerical style and formatting abilities of the computer and emphasized effects on the productivity and efficiency. Although productivity and efficiency were rarely defined in any of the studies, it was usually *assumed* (but rarely documented) that if nurses were spending less time on indirect patient care, they could use that time in more direct care and thus take care of patients more effectively.

Only recently have studies looked at the quality of the work experiences of nurses in an automated work setting (Harris 1990; Sorensen 1991). One study found that nurses felt that their sense of autonomy and control over their work had actually decreased following the introduction of computers and that there was a loss of "professionalism" (Harris 1990). In 1990, I conducted a study of a hospital that had incorporated automated nursing systems both into the day-to-day work of nursing units and into the architecture of the physical structure of the units, specifically in the form of bedside terminals. No longer were there nurses' stations but rather a space at the entrance to each patient room called the "Nurse Server." This area contained the computer terminal, telephone, paper records, medications, and treatment supplies. In this study, I found that although the nurses agreed that the computers had decreased the amount of time that they used in documenting their care, they were frustrated by the standardized documentation complaining that it often lacked the quality component of the care they had given their patients. Not only did these bedside information systems eliminate the *human quality* of their words but because of the bedside location, they often felt isolated from their peers, experienced a *decrease* in their time with patients, and actually felt that their workload had increased (Sorensen 1991). These findings are consistent with other studies that have evaluated nurses' work after automation.

It is projected that by the year 2000 only 50 percent of the needed nurses will be available to provide skilled care (Gross 1989). Although it is nowhere specifically stated that the computer will replace nursing staff, the implication has always been that with bedside terminals and the increased speed of information retrieval, *fewer* nursing staff will be needed. Today some hospitals are *actually decreasing* their nursing staff and *increasing* the unlicensed technicians to provide patient care. The new emphasis now is on what is referred to as "staff-mix," the number of assistive nursing personnel, such as nurses aides, orderlies, nursing technicians, to the number of RNs. Management information systems now focus on more accurate ways to measure the productivity of nurses with the explicit goal of reducing the numbers of nurses needed to provide the most efficient care to patients.

In this model, the unquantifiable aspects of nursing work—caring, hands-on intervention, emotional support, decision making, etc.—are increasingly squeezed out, eliminated, and unacknowledged. It is clear in the case of nursing, as so often in the history of technology in the workplace, that the introduction of automation has not produced the results originally promised. Emphasis on efficiency and productivity without equal attention to those aspects of nursing loosely grouped together under the term "caring" can seriously damage the quality both of work for nurses and health care for patients. That nurses have once again so willingly embraced the scientific model reaffirms the power of the myth of progress through science and technology, and the power of gender in defining the power of relations of role identities. There is an urgent need for a renewed emphasis on the enhancement of the unique human interactions in nursing and a more critical view of technology's ability to deliver on all its promises.

REFERENCES

Barnes, R. (1983) History of Motion and Time Study, in Hanson R. L. (Ed.), *Management Systems for Nursing Service Staffing.* Maryland: Aspen Systems Corporation.

Coser, R. L. (1963). Alienation and the social structure: a case analysis of a hospital. In Friedson E. (Ed.) *The Hospital In Modern Society.* New York: The Free Press of Glencoe.

Dunlop, M. J. (1986). Is a science of caring possible? *Journal of Advanced Nursing.* 11, 661–670.

Grier, M. R., Ziomek, R. R., MacLean, S. L. & Kim, K. K. (1985). Evaluation of a computerized nursing information system. In K. J. Hannah, E. J. Guillemin & D. N. Conklin (Eds.) *Nursing Uses of Computer and Information Science.* North-Holland: IFIP-IMIA.

Gross, M. S. (1989). The potential of professional nursing information systems. (prepared for the U.S. Department of Health and Human Services, Secretary's Commission on Nursing). Reprinted in Abrami P. (Ed.) *Proceedings from the N.E. Healthcare Assembly & Applied Management Conference on Bedside computer terminals: A concept on the crossroads?* (pp. 30–37). Boston, Ma.: N.E. Healthcare Assembly & Applied Management Systems, Inc.

Harris, B. L. (1990). Becoming deprofessionalized: One aspect of the staff nurse's perspective on computer-mediated nursing care plans, *Advances in Nursing Science. 13*(2), 63–74.

Henderson, V. (1985). The essence of nursing in high technology, *Nursing Administration Quarterly. 9*(4), 1–9.

Hendrickson G. & Kovner C. T. (1989). Effect of various components of computer systems on nurses. In Kingsland L. C. (Ed.) *Proceedings from the Thirteenth Annual Symposium on Computer Applications in Medical Care.* Washington D.C.: IEEE Computer Society Press.

Hine D. C. (1989). *Black Women in White: Racial Conflict and Cooperation in the Nursing Profession, 1890–1950.* Bloomington: Indiana University Press.

Johnson, J. E. (1988). The impact of DRGs on clinical nursing in hospital settings. In J. E. Elliot (Ed.) *Impact of DRG's on Nursing: Report of the Mid-Atlantic Regional Nursing Association.* (30–50). U.S. Department of Health & Human Services, Division of Nursing: Washington D.C.

Kjerulff, K. H. (1988). The integration of Hospital Information Systems into nursing practice: a literature review. In M. J. Ball, K. J. Hannah, U. G. Jelger & H. Peterson (Eds.) *Nursing Informatics: Where caring and technology meet.* (243–249). New York: Springer-Verlag.

Larson, P. J. (1981). *Oncology patients' and professional nurses' perceptions of important nurse caring behaviors.* unpublished doctoral dissertation, University of California, San Francisco.

Malveaux, J. & Englander, S. (1986). Race and Class in Nursing Occupations, *Sage.* Vol. III, No. 1, 41–45.

Melosh, B. (1988). "A Special Relationship": Nurses and Patients in Twentieth Century Short Stories, in A. H. Jones (Ed.), *Images of Nurses: Perspectives from History, Art, and Literature,* Philadelphia: University of Pennsylvania Press.

Noble, D. (1977). *America by Design: Science, Technology and the Rise of Corporate Capitalism.* New York: Oxford University Press.

Packer, C. L. (1986). Nurses view computers as both friends and foes, *Hospitals.* November 20, 101.

Reverby, S. M. (1987). *Ordered to care: The dilemma of American nursing, 1850–1945.* Cambridge: Cambridge University Press.

Sarason, S. B. (1985), *Caring and compassion in clinical practice.* San Francisco: Jossey-Bass Publishers.

Scott, J. W. (1988). *Gender and the Politics of History.* New York: Columbia University Press.

Sorensen, L. (1991). *Nursing and Computers: Caring in the context of information technology.* Unpublished doctoral dissertation, Graduate Center-City University of New York.
Staggers, N. (1988). Using Computers in nursing: Documented benefits & needed studies, *Computers in Nursing. 6*(4), 164–170.
Thompson, J. D. & Goldin, G. (1975). *The Hospital: A social and architectural history.* New Haven: Yale University Press.
Wajcman, J. (1991). *Feminism Confronts Technology.* University Park, Pa.: The Pennsylvania State University Press.
Watson, J. (1979). *Nursing: The philosophy and science of caring.* Boston: Little, Brown & Company.
Winner, L. (1991). Mythinformation. In J. Zerzan & A. Carnes (Eds.). *Questioning Technology: Tool, toy or tyrant?* (163–170). Philadelphia: New Society Publishers.
Zielstorff, R. (1987, April). In high-tech care, the challenge remains: to nurse, *The American Nurse.*

4
PROMOTING CRITICAL THINKING THROUGH TEACHING TECHNOHISTORY
Myra Jones

As Marshall McLuhan shrewdly noted over thirty years ago, it is much easier to understand the ideology of the technology just past than the one which still shapes the present. Current technologies—particularly those powerful enough to become the shapers of the times—are so pervasive that they disappear into the background. Just as the fish swimming in the ocean would be the last creature to discover water, most humans have difficulty understanding the implications of the current electronic communication revolution in which we are immersed. However, there is (to use McLuhan's metaphor) a "rearview mirror" in which we can view a past technology, to discover general principles which we can apply to the present. The specific assumptions, values, and details will change when the technologies change, but once the tools of analysis are learned, they can be used to identify the underlying assumptions, biases, and social effects of any technology.

Many specialists have considered these questions. Cecelia Tichi, in *Shifting Gears,* discusses the paradigm shift from the Industrial Age at the turn of the century to the Information Age, showing how the values and the metaphors changed as the technology changed. Jeremy Rifkin, in *Algeny,* examines genetic engineering's potential to become one of those technologies that change worlds and world views for good or ill. These books are useful and important, but it can be even more useful to make one's own analysis of primary sources.

This chapter will show how this type of primary source analysis can be done in group or individual projects, using the works of an American who wrote and published around the turn of the century in America. Though little studied today, Elbert Hubbard was an important technology popularizer and innovator in American society just before World War I.

From 1895 to 1915, Hubbard published a little magazine called *The Philistine,* a "Periodical of Protest." An analysis of some of the advertising and text of *The Philistine* can show Hubbard's ideas about such developments as "scientific" big business, and can reveal the implicit, as well as the explicit, attitudes of Americans toward the new technologies of that time.

BACKGROUND: ELBERT HUBBARD AND HIS HISTORICAL CONTEXT

At the turn of the century, America was undergoing a multifaceted transition: from a rural to an urban society, from an agrarian economy to an industrial one, from a culture which looked to Europe for inspiration to one that was proudly American, and from a country of regional differences to a cohesive nation. Better transportation and communication technologies were bringing the country together. It was a time of invention and new technology, a time when the old fundamentalist Protestant verities were being displaced by the new "religion" of science, a time that ushered in the increasing rate of change that has characterized this century. It was time of great optimism, when progress was a basic premise. Like most times that mark boundaries, it was also a time of contrast and conflict, between the old ideas and technologies and the new, and therefore a time which richly repays analysis.

Elbert Hubbard was a man of his times. Born in 1856, just before the Civil War, he lived in a strict fundamentalist Protestant household in the rural Midwest. Then, like many young men of his day, he emigrated to the city—first to Chicago and then to Buffalo—to make his fortune in one of the new industries, Larkin Soap Company. Hubbard got in on the ground floor of the expanding soap company as an energetic, personable travelling salesman, eventually rising to the position of vice president. Hubbard even helped to develop some of the marketing tools common today, such as premiums in the boxes of the product, and the pyramid marketing scheme made famous by such companies as Amway.

Hubbard married and began a family, and the future course of his life as a prosperous middle-class businessman seemed set. Then, in his mid-thirties, he gave it all up. He sold out his interest in Larkin to devote himself to literature and the arts. Inspired by a trip to England where he met William Morris and became enamored of the

Craft Movement, he retreated to the rural New York village of East Aurora and founded Roycroft, a craft cooperative to make beautiful things: hand-illuminated and leather-bound books, furniture, leather accessories, and other household items. Roycroft would gather like-minded artists into a commune.

Thus began the second career of Elbert Hubbard, who was to be one of the most famous and influential men of his day. Through the books and periodicals published by Roycroft and through later lecture tours, he carried his message to a wide audience.

The Philistine was one of the most widely read of the several Roycroft periodicals. Begun as a one time pamphlet in June 1895, with a printing run of only 2,500 copies, the first issue was so well received that it became a regular monthly at once, although to the end of its life, the front cover carried the slogan, "printed every little while." By 1899, circulation had grown to over 50,000 copies a month; by the time of Hubbard's death in 1915, *The Philistine* claimed a circulation of 225,000.

The Philistine had many contributors at first, including well-known writers and artists such as Stephen Crane, Rudyard Kipling, and Oz artist W. W. Denslow. But the little magazine soon became the personal outlet for Hubbard's own writing and opinion. The tone was folksy, chatty, informal, and personal, with its own spelling conventions ("enuf," "thru"). Hubbard carried on running battles with famous people, calling them snide nicknames and printing their criticisms of him. The famous evangelist Billy Sunday came in for more than his share of body punches, as did Edward Bok and the magazine Bok edited, the *Ladies Home Journal*. Hubbard used many slighting terms for this magazine, such as the "Hum Journal," the "Missus Home Journal," and the "Loidy's Own" in the April 1896 issue.

Some of Hubbard's favorite targets were the representatives of the three traditional professions: law, religion, and medicine. In this, he shows the American turn of the century faith in progress and the future and the disdain for the old, outmoded institutions that had controlled society and now resisted progress. Religion, which he considered just another superstition, was being replaced by science.

The March 1905 issue of *The Philistine* carried an essay "Theology vs. Social Science," in which Hubbard declared, "Our best knowledge of humanity is gained from observation and experience, and not from so-called Sacred Books." Calling the Church a bunch of "drones," Hubbard contrasts the Church and the State:

Our Government is a government of the people, by the people, and for the people! The Church is a government by priests who claim to be acting as Agents for Deity; and their government is of themselves, for themselves, and by themselves. The part the people play in a government by priests is to work and pay—the priests pray and play.

A typical motto in the February 1899 issue sideswipes two of this three targets: "Great preachers, like great lawyers, are sincere, but not too sincere."

Today, Elbert Hubbard is virtually unknown, and it is difficult to realize how popular and influential he was during his lifetime. His writings, particularly *The Philistine,* are a rich mother lode of information about the times. He was killed in 1915 on the *Lusitania,* at the height of his popularity, and the publication of *The Philistine* ended with his death.

THE PHILISTINE

The Philistine is a mirror reflecting not only the technology of its time but also the myths, assumptions, values, and biases which grew up around the technology. The little magazine's publisher and writer was a popularizer of many of the trends. He was an unabashed proponent of American free enterprise, and he did much to make acceptable—even praiseworthy—the growing centralization of power, and the control of the development of technology in the hands of the great corporations. Like Frederick Taylor, he helped marry the two important concepts of the new century, science and business.

The Philistine was pocket-sized, with short articles that were quick to read and easy to understand. Working people could carry it around and read it in spare free moments. It was rough brown paper with black print and red accents. The rear cover usually had a drawing, perhaps a political cartoon or a motto. There was an advertising section in front and back, and occasionally an advertising insert in the center of the magazine. Ad pages were unnumbered.

In *The Philistine,* Hubbard translated some of the most important political and philosophical ideas of the time for the growing middle class of consumer-citizens out in the American heartland. He was the cultural conduit through which farmers in Kansas, businessmen in Colorado, or ranchers in Utah kept in touch with the world. The magazine content was unrelentingly optimistic and carried the same message made popular by other writers: if you are willing to work and improve yourself, you will get ahead. Success comes with hard work. Wealth is created by the businessman who has the talent to organize people to create something new and valuable for society.

A chronological overview of the ad sections shows an evolution in both numbers of ads and types of product. Early issues have only eight pages of ads, in four-page sections front and back—with sometimes additional ads on inside front and back covers. Most of these ads are for literary products—either the Roycroft productions or others. For example, the January 1896 issue has eight pages of ads—all for literary works. These ads for literary and other cultural products reveal culture as a commodity that can be consumed.

By the following year, the products being touted in *The Philistine* had broadened, and many fall into the general category of new technology. The May 1897 issue has three pages of bicycle ads. One pictures a woman cyclist, and the text stresses the cycle's beauty. Another company has two pages, one of which appears to be directed at female customers and the other at men. The one for men begins with a literary tag: "What Shakespeare is to the poets, the Stearns Special is to bicycles." Then the text gives some technical specifications of the machine and finishes with a horse metaphor. The woman's ad features a feminine cyclist and emphasizes the cycle's looks and reliability, its "grace of outline, ease of running, staunchness, and good wearing qualities . . . just light enough to be easy-running; just heavy enough to bear all wear and tear." These cycles are not inexpensive by the standards of the day. Price of the Stearns machine (given in the men's ad but not the woman's!) is $125; the Royal retails for $100. These ads show attitudes toward technology are often gendered.

By 1898, railroad ads appeared. Hubbard eulogized trains many times both in *The Philistine* ads and in his essays. On the inside front and back covers of the January 1898 issue, we see ads for the Wabash and the Nickle Plate Railroads, both offering dining and sleeping cars. The Nickle Plate makes a special appeal to women travelling alone, saying: ". . . the day coaches are in charge of uniformed Colored Porters to render assistance to travelers, and especially to ladies and children."

Other ads appearing in 1898 reflect Hubbard's interest in a healthy lifestyle as well as the developing technology of the food industry: Postum breakfast beverage; Franklin Flour, a milled flour which proudly boasts that it is NOT a bleached flour but the more nutritious whole grain; and several health resorts, such as Dr. W. H.

Jackson's Sanitorium which contains an endorsement for another product advertised in *The Philistine,* Granola Company's health foods. Heinz and Campbell soups are also early advertisers. Hubbard said often and in print that he would only accept advertising from clients in whose products he believed. He never ran cigarette ads and he castigated another magazine whose advertising was, he declared, 73 percent cigarette ads.

By the end of 1898 the advertising section had doubled from 8 pages to 16, and the ads continue to shift from the strictly literary to the more high-tech American products new to the market. In December are ads for the Columbia Typewriter Company and Gudlach Optical Co., "the only house in the United States making all the parts that go into a camera." This shows that America is increasingly manufacturing its own high-tech merchandise, rather than importing it, and that there is a growing mass market for these products.

In December 1902 a four-page ad for Roycroft books, composed of testimonials from well-known customers indicates that Roycroft is beginning to be better known and by the right people. One letter is from the secretary of Queen Victoria, expressing the queen's pleasure in a book on Browning; another from John Ruskin's secretary retails the great man's approval of the format and typography of a Roycroft edition of *Sesame and Lilies.* The letter states, ". . . this beautiful book goes far in atoning for the typographical sins that have been inflicted on [Ruskin's] writings by certain American publishers." Other satisfied customers are actress Ellen Terry and Secretary of State John Hay.

As mentioned above, *The Philistine* reached a circulation of just under 50,000 by the end of 1898. Then, in March of 1899, came a turning point for Hubbard. In an hour he dashed off an essay that almost singlehandedly doubled *The Philistine's* circulation. The essay, "A Message to Garcia," was initially a minor piece arising, he said, from a dinner table discussion about the Spanish American War. It relates an anecdote about "a fellow by the name of Rowan" sent into the interior of Cuba to carry a message from the U.S. government to a rebel leader named Garcia. Without asking questions, Rowan "just disappeared into the jungle, and in three weeks came out on the other side of the island, having traversed a hostile country on foot, and delivered his letter to Garcia." Hubbard draws several morals from the anecdote: "It is not book-learning young men need . . . but a stiffening of the vertebra which will cause them to be loyal to a trust, to act promptly, concentrate their energies: do the thing: 'Carry a message to Garcia.'"

The little essay was immediately recognized by American corporate employers as a description of the ideal employee: loyal yet capable of initiative. The March issue of *The Philistine* sold out in three days, and reprint requests came in faster then Hubbard could fill them.

Thus started the Roycroft reprint business. Hubbard would print copies in booklet form of an essay with a company's logo on the cover, which the company could then hand out to employees or customers. One railroad bought three million copies of the Garcia essay. Hubbard, said some detractors, even wrote essays in *The Philistine* with an eye to pleasing CEOs who might buy reprints. Whether this was true or not, an ad in January 1908 directed toward "Manufacturers, Wholesalers, Department Stores, Banks, Railroads, Trust Companies, Private Schools, Colleges, and Institutions" offers to supply

> booklets and preachments by Elbert Hubbard by the thousand—your ad on the cover and a four or eight page insert, all in DeLuxe form. These pamphlets are real contributions to industrial literature. One railroad used several million. One department store used five hundred thousand. They appeal to all classes of people and are read, preserved and passed along.

There were ten titles, priced at ten cents apiece with quantity discounts.

After the Garcia essay, circulation jumped to 100,000 and thereafter climbed steadily to 225,000, a respectable figure. The circulation of only one of the most popular illustrated monthlies topped half a million: *Munsey's* with 650,000. *Cosmopolitan* and *McClure's* followed with 350,000 each, *Scribner's* with 165,000, and *Harpers* and *Century* with 150,000 each (Thomas Beer, *The Mauve Decade,* New York: Vintage, 1961, p. 173).

One of the most notable features about *The Philistine* advertising is the way the ad copy reflects the philosophy of the essays and shrewd marketing techniques. The best ads may be those that sell Roycroft's own products. Customers did not just subscribe to *The Philistine;* one became a member of the "Society of Immortals"; the society even had bylaws. When Roycroft begins selling silk ties, an ad exhorts people to buy and wear the ties so they can show that they are members of this exclusive society. An ad in August 1907 proclaims in red capital letters: "CLAIM YOUR KINSHIP . . . PROVE YOUR IDENTITY . . . BE A ROYCROFTER." Hubbard was constantly retailing stories about the workers at Roycroft, in a myth-making process that made Roycroft employees celebrities to the readers, and the readers/purchasers members of an exclusive club.

Hubbard claimed—with some validity—that his magazines were a good vehicle for advertisers. A two page ad in the July 1912 issue directed toward potential advertisers states, "When you tell your sales story thru the pages of *The Fra* and *The Philistine Magazines,* you are talking to people in sympathy with you and persuaded in your favor. Most of the advertisements, whether prepared in East Aurora or in the Big Advertising Centers, are prepared especially with this Fraternity Idea in mind. 'Special copy for the Elbert Hubbard Magazines' is the rule. . . . The readers of these magazines can be sure of the integrity of these ads, too, because when men talk out of their hearts, even thru advertising writers, they are sincere and square." Hubbard then claims that a "Big Ben" Clock ad in *The Fra* sold more clocks at a lower cost than "any other medium used by Westclox. . . ."

Many of the ads play upon the supposedly elite mental and cultural interest of the readership. An August 1907 ad for the Waltham Watch company, like many others, begins with a literary tag: "Maeterlinck's latest book is 'The measure of the Hours,' but people who are in need of a watch had better ask their horologer for 'Watch Wisdom' by Elbert Hubbard. It is given gratis. The booklet tells about Father Time & THE HOWARD WATCH, a watch that has been the standard of quality since it was first made in 1842." This booklet is one of the many Roycroft reprints. Most ad copy was written at Roycroft and reflected Hubbard's style and point of view.

One thing led to another with Roycroft and its canny owner. So many people made the pilgrimage to see "The Sage of East Aurora" that Hubbard made a marketing opportunity out of their visits, building an inn, which he thereafter advertised in *The Philistine.* As the years go on, the ads show that the daily rates go up as the accommodations get fancier. In December 1901, the Phalansterey offers meals "such as they are" for thirty-five cents, lodging at fifty cents, and weekly room and board at, "say Seven Dollars." In September 1904, the Phalanstery is described as "complete without being lavish: steam heat, running water, Turkish baths, electric lights, chapel, physician, music room, library, ballroom, potato patch and wood-pile." At this time the rates are fifty cents per day lodging and fifty cents per meal, or $2.00 per day. Later, after new buildings are built and the inn gets a more manageable name, the Roycroft Inn, rates go up to five dollars per day. As befit this new status, the language in later ads is less folksy and more stately.

The Philistine proclaimed that the Roycroft Inn was a good place for businesses and organizations to hold their conventions, and Hubbard even sponsored annual conventions of the Society of Immortals. The inn is also advertised as a honeymoon spot: An August 1914 ad brags that in June, seventy-one bridal couples have come to the Roycroft Inn. Later ads mention that garages and gasoline are available for those motoring out to the inn in August and September. Ads mention "autoists" and "auto tourists."

Hubbard also marketed this ideas and his products in his lecture tours, which were by all accounts very well attended and brought his message personally into the hinterlands.

AT&T AND *THE PHILISTINE*

The brief overview of *The Philistine* above shows that it is a comprehensive and detailed compendium of the state of technology and of American attitudes toward technology at the turn of the century. The discussion of its content also suggests many possible themes to explore.

- How technology is changing social roles, such as the role and status of women
- How manufacturing communications and transportation technologies, along with increased affluence, are changing society
  —to create a consumer culture
  —to weld the regions into a nation
- How science is becoming the new religion
  —giving rise to faith in the inevitability of progress
- How science is being applied to new areas, such as "scientific" cooking, business management, and marketing
- How big business and corporations develop and are marketed to the American people as proponents of democracy, free enterprise, progress, and patriotism
- How one particular technology evolves and increasingly influences American life, for example, the railroad, electricity, the automobile, the telephone, or food processing technology

Analysis of primary source material, in this case *The Philistine,* could take one of several approaches: use either the magazine's advertising or the text of the articles, or a combination of both. The project could either trace the development of an idea or theme (such as women's roles) in a number of ads, or could take the ads from one company and do an in-depth analysis of the evolution of content, style, values, and products.

Consider the ads of one company, American Telephone & Telegraph Co. (AT&T). This company is an especially good candidate for analysis because it was one of *The Philistine's* most faithful advertisers, so we can see the ads change over time. Secondly, the company still exists today, so we could include a comparison of current ads. Additionally, the company provided its own copy, unlike most *Philistine* advertisers, so the ads reflect AT&T's philosophy which can be compared with that of the *The Philistine*. AT&T was also one of the companies about whom Hubbard wrote an essay ("Our Telephone Service," which found its way into the reprint collection).

AT&T must have been one of *The Philistine's* most lucrative ad contracts. From the start of its relationship with the little magazine, in 1908, to its last issue in June, 1915, AT&T always took out two page ads.

The first thing a look at the ads reveals is the growing sophistication of the copy and layout. The early ads, such as the one appearing in October 1908, have very small illustrations and a great deal of copy in very small print. The October ad devotes both pages to a long argument for fair rates and against "inflexible legislative proscriptions." The December ad of that year explains in detail how the phone calls must be connected by the operator, so that customers will be courteous to the "girl" on the switchboard.

Note that the job of telephone operator has already been allocated to women and the users are assumed to be men. Another ad appearing July 1910 addressed the matter of telephone courtesy. This ad also assumes that the person making the call is a businessman and that he is discourteous to a woman operator. Someone analyzing these ads might wish to study the role of women as depicted in the company's ads. For example, an August 1914 ad notes that half of AT&T stockholders are women.

AT&T ads also chart AT&T's growth. The country's first telephone exchange, with 21 subscribers, opened on January 1, 1878, in New Haven, Connecticut. AT&T was formed in 1885. By the October 1908 issue, America already has 4 million phones. By June 1910, the company claims 5 million phones and 20 million calls per day. By September, the number of daily calls has grown to 25 million. By July 1911, there are 6 million phones in the country, and an ad the following month notes that AT&T has 120,000 employees and 75,000 stockholders. By November 1912, there are 7 million phones and over 19 1/2 million phone calls per day. By March 1914, Bell has 8 million phones and 150,000 employees.

Two interesting ads in 1914 give some idea of the magnitude of the Bell System in terms of use of materials and revenue. The March ad of that year highlights the "Unseen Forces Behind your Telephone," those materials that Bell uses: "Poles enough to build a stockade around California—12,480,000 of them worth . . . $40,000,000; wire to coil around the earth 621 times, worth about $100,000,000 . . ."(enough) lead and tin "to load 6,600 coal cars—being 659,000 pounds, worth more than $37,000,000 . . . , conduits . . . worth . . . $9,000,000 . . ., 8,000,000 (telephones worth) . . . $45,000,000 . . . , more than a thousand buildings . . . (worth) $44,000,000 . . . (and) people equal in numbers to the entire population of Wyoming—150,000 Bell System employees, not including those of connecting companies." This ad presupposes that the American public is interested in details about engineering and mechanics, which indeed it was at that time, the engineer being one of the popular heroes of dime novels and adventure stories. It also shows the American fascination with statistics.

AT&T continues to grow. The August 1914 ad entitled "How the Bell System Spends its Money" is an abbreviated annual report which reveals that AT&T has an annual gross expenditure of $215,000,000 which represents an average gross revenue of $41.75 per phone. the 150,000 employees receive $100,000,000 in wages, $11,000,000 is paid in taxes, stock and bond holders receive $47,000,000, $45,000,000 is paid for supplies, and AT&T has a healthy surplus of $11,000,000.

The AT&T ads reveal that the company has (or exploits) the American sense of progress and faith in the future; it expresses the national sense of special purpose of Americans as an example of the merits of democracy and the "melting pot" myth. AT&T claims to be a major factor in helping promote democracy and national cohesion.

An early ad, from June 1909, sets the tone. Entitled "The Sign Board of Civilization," it informs the reader that whenever the Bell logo is seen, "it stands for civilization. It is the sign of one of the most powerful influences for broadening human intelligence. . . . (It) has spread an even, highly developed civilization through the land. It has carried the newest impulses of development from town to town and from community to community. . . . Bell telephone service has brought the entire country up to the same instant of progress. *It has unified the nation.*" The ad continues in this vein, equating the Bell expansion with the "march of progress." Thus is the advancement of technology, specifically Bell's technology, equated with civilization and progress, and an American business with a sense of mission: not only making money, but also being a vital engine of democracy.

An ad in October 1910 shows a telephone user reaching out over a map of the United States to touch a city in the Midwest. The telephone is the "Annihilator of Space." In 1911, the telephone and the telegraph systems merge, and an ad in November continues the "Annihilator of Space" theme to promote the advantages of the merger to the consumer: "Already the cooperation of the Western Union and the Bell Systems has resulted in better and more economical public service. Further improvements and economies are expected, until time and distance are annihilated by the universal use of electrical transmission. . . ."

Many ads portray AT&T as being a part of the great democratic process uniting the nation. The ad from September 1910, entitled "A United Nation," pictures a long line of people using the telephone and carries the slogan, "Only by such a universal system can a nation be bound together." Another ad, from August 1913, pictures a map of the United States with a telephone at the edge of California and another on the East Coast. The states are outlined in telephone wire. This ad is a paen to American democracy:

> At this time, our country looms large on the world horizon as an example of the popular faith in the underlying principles of the Republic."
>
> We are truly one people in all that the forefathers, in their most exalted moments, meant by that phrase.
>
> In making us a homogenous people, the railroad, the telegraph, and the telephone have been important factors. The have facilitated communication. . . . bringing us closer together. . . .
>
> The telephone has played its part. . . . That it should have been planned for its present usefulness is as wonderful as that the vision of the forefathers should have beheld the nation as it is today. . . .
>
> Inspired by this need (for universal service) and repeatedly aided by new inventions and improvements, the Bell System has become the welder of the nation. It has made the continent a community.

Other ads continue the "America is better" theme and develop the idea of a central role for the phone company in protecting and extending the ideals of democracy. One ad from September 1912 has the headline "The Nile System—The Bell System." The text declares that the "Primitive makeshifts (of Ancient Egypt) have been superseded by intelligent engineering methods. . . ." building the Aswan Dam and "adapting the Nile to the needs of all the people." Like the damming of the Nile, the "same fundamental principle" operates the phone system, "intelligently guided by one policy." A 1913 ad continues the Egyptian motif, and the democracy theme, stating that the ancient Egyptian supernatural symbols of religious protection, such as amulets, have been superseded by the telephone, a real system of protection which "makes us a homogenous people and thus fosters and protects our national ideals and political rights." Thus these ads also show the Eurocentric cultural bias of "civilization" over "primitive" Egypt.

FAITH IN TECHNOLOGY

If there is any pervasive theme in *The Philistine* it is Hubbard's faith in and support of American big business. He strongly believed in its essential role as a creator and maintainer of science, progress, and democracy. An ad in the July 1915 issue of *The Philistine* for stenotype machines is typical of both his approach to ads and his philosophy.

It is in the format of an eight-page essay which is a first person testimonial, a future reprint. In this ad Hubbard expresses his mission in life, vis-à-vis business, using modern machine metaphors: "I run an idea garage and supply spare parts, lubricating oil and mental gasoline to scores of business institutions in this country. Through the medium of the *Fra* and the *Philistine* I boost Big Business and give the lie laryngismus." In this ad, one can see the distinctions between essay and ad copy blur—even disappear. Elsewhere Hubbard declares "The Science of Business is the science of supplying human wants . . ." (June 1909).

As the First World War approaches, both AT&T ads and *The Philistine* essays become more explicitly patriotic. The April 1914 issue has these words inscribed vertically up the right edge of the front cover: "A Message to Uncle Sam." A motto under the masthead gives a clue about this message: "MUCKRAKER—One who sits on the fence and defames American enterprise at it marches by."

This issue is devoted to a patriotic defense of American business and the trusts, and the charge that Germany is conducting a propaganda campaign against America, to weaken her industrial capacity.

He defends the trusts against the trust-busters and muckrakers by quoting Justice Holmes's opinion in the Chicago Board of Trade Case in favor of Social Darwinsm: ". . . the natural evolutions of a complex society are to be touched only with a very cautious hand. . . . Indiscriminate trust-busting means dynamiting prosperity and puncturing payrolls. When big business is gypped and hand cuffed, the workingman suffers." Hubbard declares that only by the efficient consolidation of smaller businesses can the American manufacturer afford the most modern industrial technology in the world.

Hubbard inveighs against the muckrakers: "Great industrial organizations should be supervised, but not by Uipton Sinclair, Lincoln Steffens, Colonel Henry, Brother Moyer, Bill Heywood, Mother Jones or Emma Goldman."

This issue has a colored center insert in glossy paper. One side pictures the American flag with the subscription "OURS"; on the opposite page is a poem, "Salute to the Flag." Tiny letters above the poem quote a newspaper account of a meeting where labor leader Bill Heywood declared "I hate the American Flag and I despise everything it stands for."

Because of Hubbard's death on the *Lusitania* on May 7, 1915, *The Philistine* discontinued publication in July 1915. However, AT&T continued to advertise with the organization's other publications. Ads in September and November 1917 show patriotic AT&T helping the war effort.

September's ad shows a large picture of a switchboard girl superimposed over an image of a city. In the bottom left and right corners of the picture are figures of Teddy Roosevelt speaking to an army general over the phone.

The text is headed "Answering the Nation's Call," and details all the ways the phone company has worked with the government. It begins by saying, "In this 'supreme test' of the nation's will, private interests must be subordinated to the Government's need." Therefore AT&T has trained 12,000 long distance operators to give precedence over "commercial messages," has doubled the phone facilities out of Washington, has installed "more than 10,000 miles of special systems . . . for the exclusive use of government departments," and has expanded telephone facilities for lighthouses, railroads, bridges, and water supply systems. AT&T has cooperated with the Navy "with brilliant success." The ad concludes by asking citizens to restrict all "extravagant and unnecessary use of the telephone."

SUMMARY

As can be seen from this short analysis of some of the AT&T ads and text in *The Philistine,* we can follow any one of a number of trails. The content is rich in imagery, myth, and allusion, as AT&T in a very self-conscious way is creating an American twentieth century vision of itself, while the *Philistine* text supports the efforts of AT&T and many other emerging businesses of the time. This type of analysis can be a very productive activity. It has the added benefit that while we learn something of history and culture, we also learn skills in analyzing media, which can then be applied to the media that dominate the present era.

5
WATERPOWER ON THE SUGAR RIVER IN NEWPORT, NEW HAMPSHIRE: A HISTORICAL AND INDUSTRIAL ARCHAEOLOGICAL INVESTIGATION
Walter Ryan

The Sugar River begins in Lake Sunapee and empties into the Connecticut River. In about twenty miles it drops just over 800 feet, some 220 of which is in Newport. The first industrial use of this water was almost contemporaneous with the advent of colonial settlement and has continued to the present day.

Newport was first settled in 1766 as part of the explosion of settlement in the upper Connecticut River Valley that took place after the fall of Quebec and the end of the French and Indian War. In the fall of that year, the Newport proprietors, most of who were from Killingworth, Connecticut, offered Benjamin Giles, of Groton, Connecticut, a mill privilege and 100 acres of land in return for his building and maintaining a gristmill and a sawmill in the town. The proprietors provided the millstones and taxed themselves the equivalent of four days labor each to help pay for the construction of the mills.[1]

Giles built his mills near the eastern border of Newport where the Sugar River cuts through Guild Hill. The location offers a natural fall of water in a place where the river was easily dammed. The remains of the foundations at this site, two walls of unmortared stone, as much as six feet thick, ten feet high in places, and forty feet long, stand twelve to fourteen feet apart right next to the river. These foundations, which may not be from Giles's original mill, could have supported a building at least twenty-four feet by forty feet and would have allowed the river to flow through the building's basement.

As the town's miller, Giles was a prominent person in the new community. The proprietors appointed him clerk in 1767 and he served the town for many years as moderator, as selectman, and as representative in the legislature.[2]

The original settlement of Newport was along a glacial terrace on the edge of the Sugar River and some three-and-one-half miles west of the site of Giles's mills. Jeremy Belknap, the author of the first history of New Hampshire, had this description of the ideal community.

> Were I to form a picture of a happy society, it would be a town consisting of a due mixture of hills, valleys and streams of water: The land well fenced and cultivated; the roads and bridges in good repair; a decent inn for the refreshment of travelers, and for public entertainments: The inhabitants mostly husbandmen; their wives and daughters domestic manufacturers; a suitable proportion of handicraft workmen, and two or three traders; a physician and a lawyer, each of whom should have a farm for his support. A clergyman of any denomination, which should be agreeable to the majority, a man of good understanding, of a candid disposition and exemplary morals; not a metaphysical nor a polemic, but a serious and practical preacher. A school master who should understand his business and teach his pupils to govern themselves. A social library, annually increasing, and under good regulation. A club of sensible men, seeking mutual improvement. A decent musical society. No intriguing politician, horse jockey, gambler or sot; but all such characters treated with contempt. Such a situation may be considered as the most favorable to social happiness of any which this world can afford.[3]

If Newport did not entirely fit this description it undoubtedly was as close to it as any other hill country town. The Sugar River runs through the town from east to west. The South Branch of the Sugar River, which rises in several small ponds south of the town, joins the main stream in what is now the center of the village of Newport. The North Branch, which drains a large area to the north, joins the main river some two and one-half miles downstream, in North Newport. One Road led from the village to the Connecticut River at North Charlestown and another from the village to Giles's mill. John Remmele was settled as the town minister. Giles, who sold his mill to Jeremiah Nettleton in 1784, served some of the functions of an attorney.[4] There was no library or musical society but there would be in the future.

Although the country was in a depression in 1790, Newport, with a population of 780 people and with steady streams of water flowing through it, was well placed to grow as the new century approached. As it did, it would, of necessity, depart from Belknap's vision.

As Newport grew, it became more connected to the outside world and more a part of the larger economy. The Croydon Turnpike opened in 1806. It passed through Newport, connecting the Second New Hampshire Turnpike in Lempster with the Fourth New Hampshire Turnpike in Lebanon, followed the east bank of the South Branch of the Sugar River and crossed the main stem of the river about a mile east of the village. Six years later the Cornish turnpike connected the Croydon Turnpike in Newport with the bridge over the Connecticut River at Windsor, Vermont.[5]

The changes that were taking place in Newport involved more than just new turnpikes. As the eighteenth turned into the nineteenth century, ambitious men built several small manufactures along both the Sugar River and the South Branch. A second grist mill, closer to the village, was built sometime in the 1780s. Nearby, the brothers Stephen and David Dexter opened a blacksmith shop sometime before 1787 where among other work they turned out scythes. About 1803, John Parmelee, who had learned his trade from the Dexters, built a scythe shop with a water-powered triphammer and grindstone on the South Branch. Reuben Bascomb soon opened a fulling mill just downstream from Parmelee, while Nathan Hurd built a carding mill and a fulling mill on the main stem of the Sugar River.[6]

With the opening of the Croydon Turnpike and the increased activity there and along the banks of the river, businesses gradually began to move down the hill and relocated near the turnpike and the river. By the early 1820s the village of Newport was ranged along the turnpike on either side of the bridge which crossed the Sugar River.

The power that drove the machinery in these early manufactures was waterpower. Very little is known about the physical configuration of most of these early mills. Except for Giles's mill, all the sites mentioned above have been built on several times. The normal waterpower engine during this period was the overshot wooden waterwheel. Most of these wheels were fourteen to fifteen feet in diameter and perhaps two and one-half feet wide.[7] Such a wheel, built around a one and one-half to two foot diameter axle, and using a wooden gear train to transmit the power from the waterwheel to the millstones or other machinery, was well within the capability of local artisans to construct.

These overshot wheels were about twice as efficient as undershot wheels, especially if the head of water was fifteen feet or more and the tailrace was sunk low enough so water did not back up and retard the turning of the wheel.[8] The dams for these mills were usually built of log cribbing, filled with stone. The upstream side of the dam was a strongback of logs covered with planks, the whole set at a forty-five degree angle to the current. To strengthen the dam the cribbing was often extended in steps on the downstream side. These steps also would be covered with planks to keep water flowing over the dam from undercutting its base.[9] A wooden flume or sluiceway led from the dam to the wheel which might often be some distance downstream.[10]

These mills were easily the most complex machines in a farming community. They worked slowly: the output of a typical gristmill might be three to four bushels of four per hour. In a community where most of the population were subsistence farmers, producing only a small surplus to sell or to barter for what they could not produce, the miller, who was allowed a toll of one-sixteenth of the grain ground, was sure to be a prominent person.[11]

Between 1813 and 1821 two major developments took place on the Sugar River that enabled manufacturers to take fuller advantage of the waterpower there.

The first was a power canal in Newport Village. James Walcott, an entrepreneur from Rhode Island, moved to Newport about 1812 and in concert with William Cheney bought about forty acres of land along the Sugar River just to the east of the Croyden Turnpike. They then dug a power canal almost 400 yards long. The canal, which was on the north side of the river, drew its water from an existing dam and provided power for a cotton spinning mill which Walcott and Cheney built.[12] The two men then sold off mill lots and water rights to others. Soon, local entrepreneurs built a tannery and a linseed oil mill beside the canal.[13]

The second development was the construction of a dam at the outlet of Lake Sunapee. Mill owners along the Sugar River, led by Josiah Stevens, petitioned the New Hampshire Legislature for incorporation and for authorization to dam the outlet of Lake Sunapee. Their purpose was to better control the water in the Sugar River. (New Hampshire had no general law of incorporation at this time. Corporations were chartered by individual acts, passed by the legislature and signed by the governor.)

The act of incorporation was approved December 7, 1820. The Sunapee Dam Corporation could sink the outlet of the lake ten feet below the low water mark, build a dam and gates at the outlet, and control the water for the benefit of the mills along the Sugar River. The corporation, which would be managed by three directors selected from among the owners of the mills and mill privileges along the Sugar River, could assess mill owners along the river for funds to build and maintain the dam and to pay damages to any person whose property was injured because of the operation of the dam.[14] Cheney, who was the owner of the land at the outlet of Lake Sunapee, as well as being a mill owner, granted the corporation the right to build on and to pass over his property. In return, Cheney and Walcott's mill would always be free from any assessments made by the corporation.[15] The corporation organized itself on June 26, 1821, and soon completed the dam.

There were other, smaller, efforts to channel the waters of the Sugar River during the nineteenth century. In 1827, Israel Kelly built a loose boulder dam and a 2,000 foot long canal to power his sawmill. Several years later, Samuel Larned built a dam and canal in North Newport to provide power for his scythe shop.[16] What was probably the last such effort took place in 1892, when W. T. Millekin dug a canal beside the Sugar River just west of Main Street in downtown Newport and built a log boom in the river to channel the water into it. Millekin had built a new building that housed a machine shop on the lower floor and a woodworking shop on the upper. His canal, which had an eight foot head of water, provided the power for both shops.[17]

By the early years of the nineteenth century, many people, particularly mill owners, saw water almost solely in terms of its economic value. The mill owners felt a need to control both the supply and the flow of the water that powered their mills. Lake Sunapee was neither the only nor the largest lake that was dammed to provide power for mills; almost all of New Hampshire's lakes were used in that manner, starting with the largest, Lake Winnipesaukee.[18] Part of the desire of the mill owners for increased control over their waterpower was the fear that as farmers and loggers cleared more land, stream flow was becoming more irregular, with higher freshets in the spring and longer droughts in the summer.[19]

Although mill owners in Newport, throughout New Hampshire, and in the Northeast generally, were making great efforts to expand their waterpower, they were not unaware of the possibilities of steam power. The first steam engine in what is now the United States was erected in 1754 to pump water out of a copper mine in New Jersey.[20] What was probably the first steam engine in New England was in place in Pawtucket, Rhode Island, by 1811.[21] Steam engines did not come to New Hampshire, however, until 1833. By 1838, the year in which the railroad entered New Hampshire, there were six stationary steam engines and one steamboat in the state.[22]

While work on steam engines was taking place, however, there also was work being done on increasing the efficiency of waterpower. As long as many machines in a factory are powered by a single engine, it makes very little difference in the energy being delivered whether the engine is powered by water or steam. If waterpower is affected by low water or by freshets, steam is dependent on an adequate supply of fuel. Factory owners in New England, where there is an abundance of waterpower sites and a relative scarcity of fuel, preferred to increase the efficiency of their waterpower rather than convert to steam.[23]

Developments in waterwheels went forward on two different lines. Traditional waterwheels began to be constructed out of iron. An iron waterwheel, as opposed to a wooden one, could be made larger, both in diameter and width, was stronger, and was lighter for its size. The largest iron wheels, which reached diameters of eighty feet and widths of fifteen feet, achieved efficiencies of as much as 88 percent.[24] As these wheels became larger, overshot wheels gave way to breast wheels. Although for a given diameter of wheel a breast wheel was less efficient than an overshot wheel, the size of an overshot wheel is limited by the height of the head of water. The breast wheel, on the other hand, can have a diameter considerably larger than the height of the head of water. Because the torque produced by a waterwheel is dependent on the diameter of the wheel as well as on the force of the water, a breast wheel can often provide more power than can an overshot wheel for a given waterpower site. In the United States, breast wheels, which were uncommon in earlier years, were well known by the 1830s.[25]

The other major development in waterpower technology started with the tub wheel and culminated in the development of the hydraulic turbine. A tub wheel is a horizontal waterwheel on a vertical shaft. The wheel is set within a circular wall or tub which directs the water through the wheel. The impact or force of the water hitting the vanes of the wheel provides the power. Although a tub wheel is not very efficient, it can make use of a very low head of water.

A reaction wheel is similar to a tub wheel. Reaction wheels, however, are turned by the pressure, rather than the force, of the water on the vanes of the wheel.[26] The most visible difference between a tub wheel and a reaction wheel is that the vanes or buckets in the reaction wheel are curved. The water flows into a reaction wheel from above, passes through the curved buckets, and runs out along the bottom periphery of the wheel.

Local artisans often built both tub wheels and reaction wheels. The development of the tub wheel into the reaction wheel was the result of many of these self-taught mechanics proceeding on a trial and error basis. Benjamin Tyler of Lebanon, New Hampshire, was one of them. He patented what he called a "wry-fly" wheel in 1804. This was a reaction wheel but it anticipated some of the principal features of the hydraulic turbine.[27]

Reaction wheels, which were in common use by the late 1820s, averaged 40 percent efficiency.[28] They were less efficient than a breast wheel, but had two major advantages: they could be used with a much lower head of water and they could be run submerged.[29] Reaction wheels, with their curved buckets, did require knowledge of metalworking but their construction was well within the capabilities of local blacksmiths and millwrights.

Between 1829 and 1840 Zebulon and Austin Parker, working in Ohio, patented two inventions that transformed the reaction wheel into the hydraulic turbine. While experimenting with the reaction wheel that powered their mill, they developed what they called the "helical sluice." This is a spiral casing which channels the water entering the wheel into the line of the wheel's motion. Reaction wheels with helical sluices were close to 65 percent efficient. The Parkers' second invention was the draft tube. About 1831, Austin Parker realized that either the pressure of the water from above or the suction of the water from below would cause the wheel to rotate. A draft tube is an airtight and watertight casing through which the water flows into the tailrace after it has passed through the wheel. The use of a draft tube allowed reaction wheels to be placed above the level of the tailrace with no loss of head.[30]

Shortly after the draft tube was patented, Elwood Morris imported a Fourneyron turbine from France which he described in the *Journal of the Franklin Institute*. At about the same time, Uriah Boyden, working in Lowell, Massachusetts, made further improvements in the Fourneyron turbine which decreased the power lost due to turbulence as the water passed through the turbine. Boyden patented his improvements and sold them to the Lowell Corporation in Massachusetts in 1849. Boyden joined James B. Francis, the Lowell Corporation's engineer, and together they continued their work on turbines. This is the stage at which the main work on turbine development passed from the millwright to the engineer.

Francis had been working for some years on a series of experiments at Lowell designed to decrease both friction and water resistance in turbines. Boyden and Francis combined mathematical analysis, graphical analysis, and experimentation to develop increasingly efficient turbines. Their approach was to direct the water smoothly into the turbine and to have it leave the turbine with minimum velocity, thus using all of the energy of the water to turn the turbine. As the two men worked they reduced the number of buckets while making them longer. This decreased the diameter of the turbine and lengthened the axle. These characteristics allowed the turbine to turn at high speed. Working with a thirty-six inch diameter turbine and a twenty-six foot head of water, they were generating fifty-five horsepower by 1855 when they published their experiments. Other experimenters, building on their work, increased the output to ninety-six horsepower by 1860 and to 266 horsepower by 1876.[31]

Just as steam did not immediately replace waterpower, turbines did not immediately displace the older waterwheels. Still, by the 1850s, turbines were common in the larger mills.[32]

In Newport, the 1840 census showed a population of just under 2,000 people, just over 4,000 sheep, and more than 800 pounds of wool. There were two fulling mills and one woolen mill employing a total of eight people, two tanneries that employed six people, and two grist mills and ten saw mills employing a total of ten workers.[33] Ten years later, although the town had only grown by sixty-two people, 790 people were employed in manufacturing.[34] Although 1840 was a depression year and 1850 was not, something more seems to be going on. Newport was changing from a farming town to a mill town.[35] In the intervening ten years two new textile mills had been built in Newport. Perley S. Coffin and John Puffer built the Sugar River Mills shortly after 1840. Thomas Twitchell and Royal Parks, who in 1838 bought the mill owned by the Newport Manufacturing Company when that company failed, were making cassimeres, a fine, twilled woolen cloth used for men's clothes, in 1840. Twitchell bought out Parks a few years later and added broadcloth, tweeds, and flannel to his production. Parks and Oshea Ingram then bought an old linseed oil mill and refurbished it to produce cassimeres.[36] In addition to these textile mills Ezra Sibley and William Dunton were manufacturing scythes in North Newport and several artisans were turning out wood products from handles for farm tools to fine furniture in shops along the Sugar River and on several of the small brooks that flowed into it. These, together with several grist mills and saw mills, produced the products that provided Newport's economic base.[37]

In spite of this impressive total, the reality was that these firms were small and undercapitalized. The smallest of the sixteen manufacturers using waterpower was Leander Long. He used waterpowered machinery in his woodworking ship where he had 400 dollars of capital invested, and employed two men whom he paid a total of fifty-two dollars a month. Working together, in 1850, they made 1,000 dollars worth of bedsteads. Twitchell, the largest manufacturer in the town, had invested 20,000 dollars in his mill where he employed fourteen men and seven women to produce 179,000 yards of cloth a year. The average wage for the men was just over twenty dollars a month while that of the women was a little more than fourteen. The cloth sold for about thirty cents a yard.[38]

Newport grew, but grew slowly, during the next few decades. The telegraph reached Newport in 1866, the telephone in 1883, and electric lights in 1892. Perley S. Coffin and William Nourse built a new woolen mill in 1867 near the site of Benjamin Giles's original mill.[39] The railroad connected Newport to Concord in 1871 and was pushed through to Claremont in 1872. Huntoon and Ladd had a steam boiler in their tannery by 1871 and in 1877, Frank P. Rowell built a steam-powered grist mill, which could grind a bushel of corn a minute, near the center of the village and away from the river.[40] Although steam engines were in common use in New Hampshire, as late as 1880 only one of the sixty-eight manufacturers using waterpower on the Sugar River and its tributaries, one of the three paper companies in the neighboring town of Claremont, had a supplementary steam engine.[41]

The publication of the 1870 census figures led the editor of the local newspaper, the *Argus and Spectator,* to point out that in Sullivan County only the towns along the Sugar River had increased in population. What he called "back farms," those hill country farms away from the river, were being abandoned as the people moved into the villages.[42]

The early 1870s were a period of general prosperity. Coffin and Nourse's Granite State Mills employed over fifty workers who produced 550,000 yards of fabric a year on four carding machines, three spinning jacks with a total of 236 spindles, and twenty looms. Needing more power, they removed a Tyler patent turbine and replaced it with a 250-horsepower Cooks Giant Turbine. Ezra Sibley enlarged his scythe factor in North Newport and installed new machinery. Dexter Richards and his son Seth, owners of the Sugar River Mills, demolished their old building in 1873 and built a modern four story stone, iron, and brick mill building with a tinned roof.[43] (The present

Figure 5.1 Looking northwest toward Richard's Sugar River Mill as it appeared between 1873 and 1905. The tracks of The Claremont and Concord Railroad run close to the river here. Sunapee Street is in the background. (With permission of the New Hampshire Historical Society.)

Figure 5.2 This map shows the business and residential section of Newport, New Hampshire, in 1892. Richards's Mill is between Sunapee Street and the Sugar River on the east side of the village. The South Branch joins the Sugar River just north of Elm Street. (With permission of the New Hampshire Historical Society.)

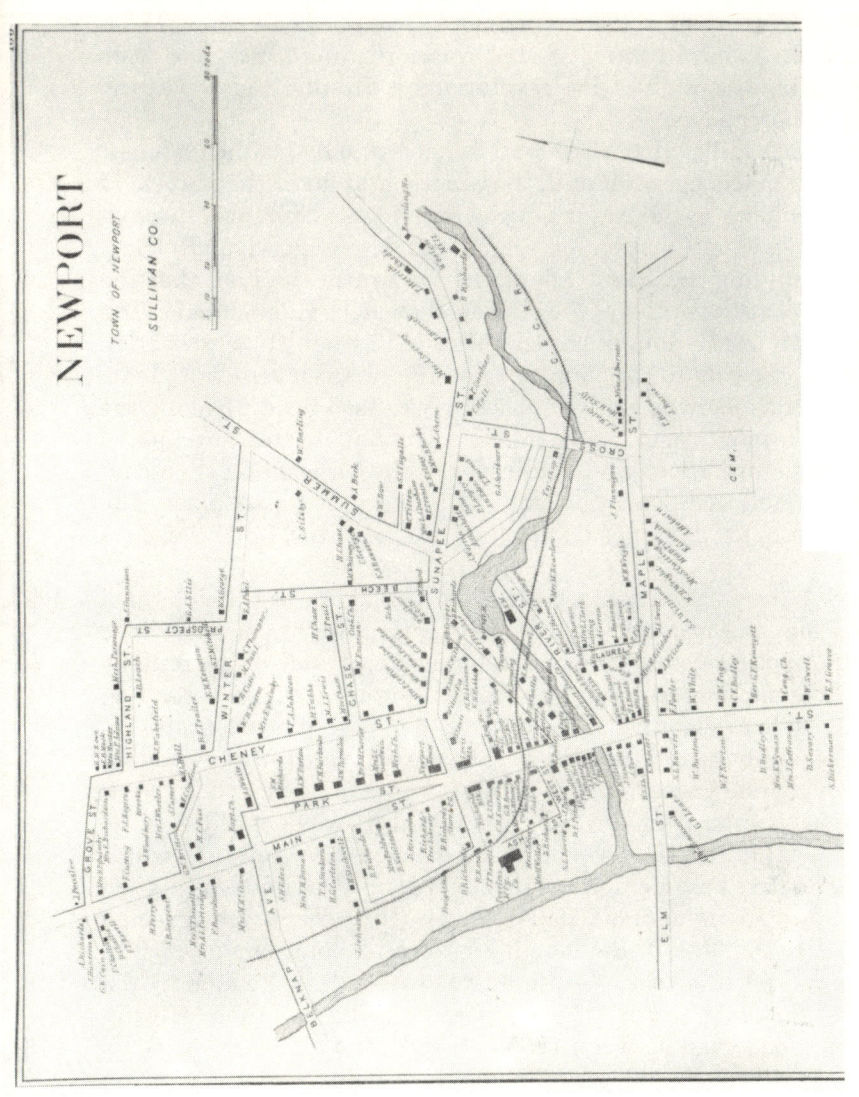

mill building, a four story brick building with a flat roof and a tall, square Italian Romanesque bell tower at the northwest corner of the building, dates from 1905.) The Richards' employed thirty-two men, thirty women, and twenty-five children who worked eleven hours a day spinning thread and weaving cloth. The power to run the machines came from a ten foot fall of water turning a five foot wide Tyler Turbine at eighty five revolutions a minute and producing sixty-five horsepower.[44]

All of these mills, all of their waterwheels and hydraulic turbines, all of their machinery, all of their owners, and all of their workers were dependent on the water flowing out of Lake Sunapee, through the Sunapee Dam Corporation's dam, and down the Sugar River.

When building the original dam, the corporation had not changed the channel at the outlet of the lake as allowed by their charter. The first change in the dam occurred about 1845 when the corporation added flashboards to the dam to raise the water level in the lake. When the flashboards were in place they caused flooding along the land of lakeside property owners. The original charter anticipated that this might happen and required that the corporation should offer "reasonable compensation" for any damage that took place. The first requests for damages do not show up, however, until the early 1850s.

In 1851 the corporation rebuilt their dam, lowering the channel two and one-half feet, and clearing obstructions to the channel above the dam. Workers sunk timber uprights on each side of the outflow of the lake and bound them together with other timbers extending from one side to the other. Planks were then placed horizontally between the timbers and the level of the lake was controlled by adding or removing one or more of the planks.

It appears that after the dam was rebuilt it was possible for the corporation to raise the level of the lake above the usual high water mark. The first suit for flowage involved John Pike and was settled in 1855 by a payment of 300 dollars and by the corporation agreeing not to raise the level of the lake higher than a point one foot below an iron pin set in a rock. This level was marked on the dam by the numeral ten on the gauge there. There were other claims for damages due to flowage between 1855 and 1861 and all were settled by payments of between 50 and 300 dollars.

The dam was reconstructed in 1870 and shortly after, a gate was built into the dam so that the water level could be controlled without having to add or remove planks. In neither instance was the height of the dam changed.

About 1875, the character of Lake Sunapee started to change. Summer vacationers started to be attracted to the area and by 1882, summer cottages and lakeshore hotels and boarding houses were common around the lake. The *Lady Woodsum,* a 100-passenger steamer plied the lake, meeting the trains and providing transportation to several lakeshore landings.

The year 1881 was very dry. The corporation mismanaged the water supply in such a way that there was not enough water for either the mill owners, for safe navigation, or for vacationers. These problems were serious enough so that in 1886 the corporation revised their bylaws. They specified that of the three directors, one would have to reside in Sunapee, one in Newport, and one in Claremont and that the object in controlling the gates was to equalize the flow of water in the river as much as possible. The corporation hired a Mr. Flanders, who lived in Sunapee, to be the resident manager of the dam. Although Flanders was able to maintain a relatively even water level, there continued to be problems associated with the level of the lake.

The corporation in 1896 rescinded the requirement that the directors come from the three towns through which the Sugar River flowed. That same year, Flanders was replaced by a Mr. Abbott. Abbott was not only the resident manager of the dam he was also an employee of the pulp mill that operated in Sunapee Harbor.

Once again, the dam was mismanaged. The spring of 1897 was rainy and the water was high but Abbott kept flashboards on the dam until June 12 when, having received complaints, the directors ordered him to lower the level of the lake. By this time wharves were submerged, cellars were flooded, beaches were damaged, and the state fish hatchery was injured. At other times Abbott had drawn the water so low that the exposed shore line extended for several rods in some places.[45] This caused the shoreline to be less than attractive, and navigation to be dangerous as well as cutting the supply of water for the fish hatchery.

Part of the problem was that increased demands on the water of the lake were incompatible. Even with the best management they might not have been able to have been met. Electric generating plants, because they operated twenty-four hours a day, used more water than a mill with a similar size turbine. The pulp mill in Sunapee also ran day and night. In addition, the owners of the pulp mill accumulated logs in the lake and then, using additional water, sluiced them through the dam and down to their mill some distance beyond. As a result of Abbott's misfeasance and the Sunapee Dam

Figure 5.3 Sunapee Harbor, c. 1890, with two of the steamboats that plied the lake. The tannery, paper company, and other industries shown on the map of Sunapee Harbor are out of view to the left of this photograph. (With permission of the New Hampshire Historical Society.)

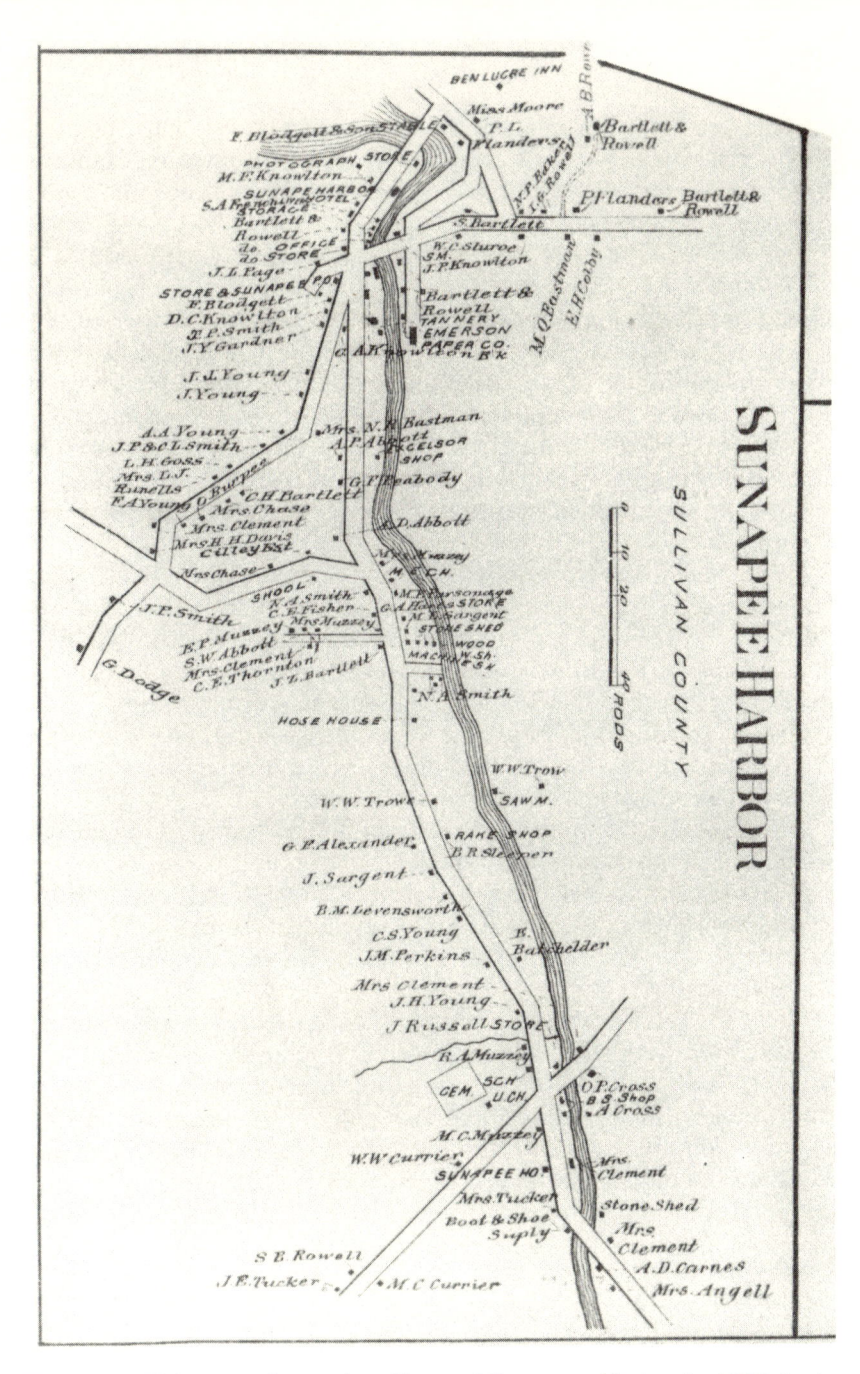

Figure 5.4 This map shows the village of Sunapee Harbor in 1892, looking south. A small section of Lake Sunapee is shown at the east end of the village. The dam is right next to the road where the lake drains into the Sugar River. (With permission of the New Hampshire Historical Society.)

Corporation directors' neglect, the corporation was sued by the State of New Hampshire, the Woodsum Steamboat Company, Hopkins, Hay[46], Quackenbos, Dunning, Bradley, and Fitch. These last were all owners of lakeshore property.

The court found that Abbott had not managed the dam with reasonable care. Beyond that, they found that the corporation should be able to manage the dam in a reasonable manner so that the mills would have the necessary water and the rights of the owners of the shoreline, of navigators, and of anglers would not be seriously impaired. The court also found that reasonable management of the dam would result in a more stable water level than would happen if the lake were left in its natural state. Finally, the court found that running logs through the dam in the summer was not reasonable.

The period of waterpower was coming to an end. In the mills that drew power from the Sugar River, lumber, pulp, paper, frames, clothespins, machinery, paint, monuments, rakes, electricity, drills, cotton, woolen, and linen textiles, shoes, flour, scythes, and boxes were manufactured. These mills represented several million dollars of capital, and about 2,000 people were employed in them. They were the economic mainstay of Newport and Claremont. But a new industry, tourism, also had claims on the water.

The investment in tourist hotels, summer cottages, and lakeside residences in the three towns which border Lake Sunapee, New London, Sunapee, and Newbury, was over 500,000 dollars. In addition a fish hatchery and four steamboats used the waters of the lake. As the court noted, the value of this property was dependent on a stable water level.[47]

Although waterpower was being replaced first by steam and then by electricity, both the lake and the river remained economically important. During the 1930s as more and more of the mills began to use electricity to drive their machinery, they dropped out of the corporation. Finally, in 1961, the state took over the dam. The two woolen mills then remaining in Newport and the paper mills downstream in Claremont used water from the river in the processing of the cloth and the manufacture of paper. The state's first concern, however, was to manage the dam to accommodate the owners of the shoreline around Lake Sunapee.[48]

Newport, Sunapee, and Claremont, the three communities through which the Sugar River flows, were settled by pioneers of English stock before the American Revolution. In the nineteenth century, Newport and Claremont grew into what we now think of as typical New England mill towns. Ethnic groups other than English became part of both communities.[49] Gradually, agriculture became a less important part of these communities' economic base.

In the latter part of the nineteenth century, Sunapee, which was never industrialized, became a destination for tourists and vacationers who were drawn to the recreational opportunities offered by the waters of Lake Sunapee. All three communities are still identified by their nineteenth century transformations. The urban centers of both Newport and Claremont are characterized by nineteenth and early twentieth century red brick industrial, commercial, and governmental buildings. Neither presents the traditional facade of a white clapboarded New England village centered around a town common. The economy of Sunapee is based on the recreational opportunities afforded by the lake and by nearby ski areas.

At the present time, the Dorr Woolen Company still uses river water in the processing of its cloth, and the dam and turbine at the Richards Mill have been refurbished to generate electricity. Although most manufacturers in Newport are still sited near the river, all other use of the Sugar River in Newport is recreational.

NOTES

1. *Newport Town Records, I, Proprieter's Records,* 20, Town Office, Newport, New Hampshire.

2. *Newport Town Records, I, Proprieters Records,* 22. *Newport Town Records, Book A,* passim, Town Office, Newport, New Hampshire.

3. Jeremy Belknap, *The History of New Hampshire,* (Boston, 1813), vol. III, p. 251.

4. Vol. 15, p. 443, Cheshire County Registry of Deeds. From an analysis of the penmanship of wills in the Cheshire County Probate records, it is evident that Giles wrote several wills for residents of Newport.

5. Joseph W. Parmelee, "History of Newport," in D. Hamilton Hurd, ed., *History of Cheshire and Sullivan Counties, New Hampshire,* (Philadelphia, 1886), pp. 218–20. Brian D. Burford, "Early Turnpike Roads in New Hampshire," *The Benchmark,* New Hampshire Land Surveyors Association, vol. 2, no. 4, Dec. 1978, pp. 41–2. Sam H. Edes, *Tales From the History of Newport* (The Argus Champion, n.p., n.d.), pp. 25–6.

6. Parmelee, pp. 218–25. James L. and Donna-Belle Garvin, *Instruments of Change, New Hampshire Hand Tools and Their Mak-*

ers, 1800–1900, (New Hampshire Historical Society), p. 71 and 85. Fulling is a cloth finishing process in which the woven cloth is washed and beaten in order to shrink and thicken it. Carding is the process of combing textile fibers such as cotton or wool to align them and prepare them for spinning.

7. Terry S. Reynolds, *Stronger Than A Hundred Men, A History of the Vertical Water Wheel,* (The Johns Hopkins University Press, Baltimore and London: n.d.), pp. 306–10.

8. Louis C. Hunter, "Waterpower in the Century of the Steam Engine," in *America's Wooden Age: Aspects of its Early Technology,* Brooke Hindle, ed., (Sleepy Hollow Restorations, Tarrytown, N.Y., n.d.), p. 67. The theoretical output of a waterwheel may be computed by multiplying the weight of the water in the buckets by the radius of the wheel. Hunter reports that efficiencies of 50 to 70 percent were standard for overshot wheels.

9. For a description of a timber crib dam of a somewhat later date see David R. Starbuck, "The Timber Crib Dam at Sewall's Falls," in *The Journal of the Society for Industrial Archeology,* vol. 16, no. 2, 1990.

10. Charles Howell, "Colonial Watermills in the Wooden Age", in *America's Wooden Age,* Hindle, pp. 134–5. The remains of two dams in Newport, both of which were in use into the twentieth century, exhibit this type of construction. See also Oliver Evans, *The Young Mill-Wright's and Miller's Guide,* in any one of several editions for information on contemporary practice in the construction of waterwheels, power trains, and dams.

11. Howell, pp. 127–8, 157–9. The one-sixteenth is specified in a New Hampshire Act approved December 9, 1797.

12. Parmelee, pp. 225–6. Richard Parker, "The Mystery of Canal Street," *Newsletter, Society for Industrial Archeology, New England Chapters,* vol. 12, no. 2, 1992, pp. 9, 10.

13. Parker, pp. 9, 10.

14. State et als v. Sunapee Dam Co., Report of Hon. John M. Mitchell, Referee.

15. Deed, William Cheney to Sunapee Dam Corporation, June 2, 1821, vol. 89, p. 139, Cheshire county registry of Deeds.

16. Edmund Wheeler, *History of Newport, N.H. from 1766 to 1787,* (Concord, 1879), pp. 74–5.

17. *The Republican Champion,* July 28, 1892.

18. Theodore Steinberg, *Nature Incorporated, Industrialization and the Waters of New England,* (Cambridge University Press, 1991) is a discussion of the relationship between water as a natural

resource and the industrialization of New England particularly as it was played out along the Merrimack River.

19. William Cronon, *Changes in the Land,* (Hill and Wang, New York, 1983), makes the point that deforestation led to irregular stream flow. This point also is made by Charles D. Smith, "A Mountain Lover Mourns, Origins of the Movement for a White Mountain National Forest 1880–1903, *The New England Quarterly,* March, 1960, vol. 33, pp. 37–56 and, in passing, by James Wright, *The Progressive Yankees, Republican Reformers in New Hampshire, 1906–1916,* (University Press of New England, Hanover and London, 1987.) Jamie T. Eves, " 'Shrunk to a Compative Rivulet': Deforestation, Stream Flow, and Rural Milling in 19th Century Maine," *Technology and Culture,* (vol. 33, 1992), pp. 38–65, argues that ponds and bogs served as natural stream flow regulators and that deforestation resulted in few problems for millers.

20. Louis C. Hunter, *A History of Industrial Power in The United States 1780–1930, vol. 2: Steampower,* (published for the Hagley Museum and Library by the University Press of Virginia, Charlottesville, 1985), pp. 1, 3.

21. Carroll W. Pursell, *Early Stationary Steam Engines in America, A Study in the Migration of Technology,* (Smithsonian Institution Press, City of Washington, 1969), pp. 83–5.

22. Hunter, "Waterpower in the Century of the Steam Engine," pp. 191–2. Purcell, pp. 73–4.

23. Pursell, p. 85. Hunter, *A History of Industrial Power in the United States 1780–1930, vol. 2: Steam Power,* pp. 84–7.

24. Reynolds, pp. 306–10.

25. Hunter, "Waterpower in the Century of the Steam Engine," pp. 67–71.

26. Force and pressure are used here in their technical sense. Force is a push and is usually measured in pounds. Pressure is force per unit of area and is usually measured in pounds per square inch.

27. Edwin T. Layton, Jr., "Scientific Technology, 1845–1900: The Hydraulic Turbine and the Origins of American Industrial Research," *Technology and Culture,* (vol 20, 1979), p. 65.

28. Layton, pp. 67–8.

29. Hunter, "Waterpower in the Century of the Steam Engine," pp. 299–303.

30. Layton, pp. 687–9. Hunter, "Waterpower in the Century of the Steam Engine," pp. 308–10.

31. Layton, pp. 64–87.

32. Hunter, "Waterpower in the Century of the Steam Engine," pp. 184–5.

33. *Statistics of the United States of America–Sixth Census,* (Washington, 1841), pp. 34–5.

34. *Statistical View of the United States, Being a Compendium of the Seventh Census,* (Normal Ross Publishing Inc., 1990), pp. 226–7.

35. Harold Underwood Faulkner, *American Economic History,* 8 ed., (Harper & Row), pp. 640–1.

36. Parmelee, "History of Newport," pp. 225–6.

37. Henry E. Mahony, ed., *Newport, New Hampshire, 1761–1961, Bicentennial Booklet,* (Newport, 1961), p. 53. For a list of Newport artisans, see James L. and Donna-Belle Garvin, *Instruments of Change, New Hampshire hand tools and their makers, 1800–1900,* (for the New Hampshire Historical Society by Phoenix Publishing Canaan, New Hampshire, 1985), passim. For furniture makers in Newport see Donna-Belle Garvin, "A 'Neat and Lively Aspect': Newport, New Hampshire as a cabinetmaking Center," *Historical New Hampshire,* (vol. 43, no. 3, Fall, 1988). For the value of manufacturing production see *Statistical View of the United States,* pp. 226–77.

38. *Mss Census Records, 1850,* New Hampshire State Archives, Concord, New Hampshire.

39. This is now the Dorr Wollen Company, owned by Pendleton Woolen Mills. It is the only textile mill in Newport.

40. The steam engines are mentioned in the *Argus and Spectator,* December 15, 1871, October 12, 1877, and November 23, 1877. For dates on the railroad, the telegraph, and the telephone, see Parmelee, "History of Newport," pp. 227–8. The date on electricity is from the *Argus and Spectator,* June 3, 1892.

41. *Report on the Water-power of the United States, Part I,* (Washington, Government Printing Office, 1885), p. 127.

42. *Argus and Spectator,* October 18, 1872. For a discussion of the changing economic and agricultural climate in New England see Harold Fisher Wilson, *The Hill Country of Northern New England, Its Social and Economic History, 1790–1930,* (AMS Press, Inc, New York, 1967) and Howard S. Russell, *A Long, Deep Furrow, Three Centuries of Farming in New England,* (University Press of New England, Hanover, New Hampshire, 1976). *The Hill Country of Northern New England* is a reprint of the Columbia University Press, 1936 edition.

43. *Argus and Spectator,* January 5, 1872, May 24, 1872, May 23, 1873, August 1, 1873. See Bryant F. Tolles, Jr., with Carolyn K. Tolles, *New Hampshire Architecture, an Illustrated Guide,* (Univer-

sity Press of New England, Hanover, N.H., 1979) for a description of this building.

44. *Mss Census Records, 1880, Schedule 3, Manufacturers,* New Hampshire State Archives, Concord, New Hampshire.

45. A rod is a unit of length equal to sixteen and one-half feet. It was commonly used to measure land.

46. John M. Hay, who owned a lakeside estate in Newbury, served as Abraham Lincoln's private secretary and in later years was ambassador to Great Britain and United States secretary of state.

47. The account of the controversy surrounding the control of the dam is based on *State et als v. Sunapee Dam Co.,* Report of Hon. John M. Mitchell, Referee, pp 1–17.

48. State of New Hampshire Environmental Services, Department of Water Resources, file 229.04.

49. Dexter S. Richards, Newport, New Hampshire, Dec. 2, 1919, letter to Lizzie M. Richards (his mother Elizabeth), Newport, New Hampshire, Historical Society. He lists the ethnicity of the workforce as follows: Americans: 92, Irish: 12, all born on this side except one, making total Americans: 104, Finns: 44, French: 48, Polish: 18, Greeks: 10, Italians: 3, Austrians: 3, Germans: 1. "Out of this total of 231 there are approximately 75 who are not naturalized." For a discussion of ethnicity in Claremont see Kathryn Grover, "The Orthodox Russians of Claremont, New Hampshire," *Historical New Hampshire,* (vol XXXIV, no. 2, Summer, 1979). William L. Taylor, "The Nineteenth Century Hill Town: Images and Reality," *Historical New Hampshire,* (vol XXXVII, no. 4, Winter, 1982), notes that immigrant Irish laborers who worked on railroad construction crews in New Hampshire moved on with the construction of the railroad and therefore had little long lasting impact on the ethnic makeup of the towns through which the railroad passed.

6
ELECTRICAL COMMUNICATION, LANGUAGE, AND SELF
David Hochfelder

The year 1994 marks the 150th anniversary of commercial telegraphy in the United States. In this period American historians have largely failed to confront the cognitive and linguistic effects of electrical communication devices. This oversight is surprising since Americans of the previous century recognized that these technologies constituted a sharp break with past forms of communication. The telegraph and the telephone not only increased the speed with which Americans communicated; more importantly, they encouraged Americans to conceive of communication in increasingly abstract ways.

This abstraction took three forms. First, business uses of the telegraph fostered an ethos of "control through communications."[1] Commodity traders in the midwest grain trade after 1850, as will be discussed later in this chapter, employed the telegraph to establish spatial and temporal control over markets. Moreover, telegraphy compressed and standardized the written language, encouraging its transformation into a medium designed for efficient information transfer. Finally, the technical and economic imperatives of the telegraph and the telephone led, if not caused, engineers to conceive of communication as a depersonalized, mathematical process. Innovations in multiple telegraphy and long-distance telephony illustrate this conception.

Communication engineers in the 1940s wove these abstractions into their discipline's founding assumptions. For them communication was inseparable from control over physical processes. They labeled any information which did not meet this purpose as "noise": irrelevant at best, destructive at worst. Since control requires efficient information transfer, communication specialists defined efficiency as their primary object. To this end they treated language as an imprecise string of symbols containing redundancies, distortions, and gaps. They regarded the human as a source of ambiguity and unpredictability in the communication link. The mathematization of communication, which began with innovations in multiple telegraphy and long-distance telephony, legitimated these assumptions. Communication became the transmission of an arbitrary message through an abstract medium; engineers assumed out of existence the message content and the communicators.

This process of mathematization culminated in the postwar disciplines of cybernetics and information theory. Its founder Norbert Wiener intended cybernetics to become "the entire field of control and communication theory, whether in the machine or in the animal."[2] Information theory, closely related, focused on computer-based information processing.[3] Mathematical models of communication and information authorized researchers in these fields to blur from both sides the distinction between human and machine. Cyberneticists borrowed feedback and process control concepts from electrical engineering to describe human physiology and psychology. Information theorists appropriated linguistic competence and cognitive ability, fundamental human qualities, to design early computer systems.

The abstraction of communication penetrated beyond the boundaries of these disciplines. The ideology of control through communication shaped the first half-century of telephone use; Bell System managers before 1920 marketed the telephone as a businessman's instrument or as an aid to middle-class household management. The early rate structure, a flat monthly fee regardless of the number of calls made or their duration, assumed that calls were brief exchanges of information and not extended conversations. The introduction of measured rates, assessed per call, marked the end of the Bell System's efforts to restrict the telephone to functional uses. The ethos of control helped to maintain traditional gender roles, as men used the telephone chiefly for business purposes and women "visited" over it.

The assumptions at the heart of electrical communication also influenced conceptions of self-identity. Role-theory sociologists, especially Erving Goffman,[4] described social interaction as a feedback-control process in which individuals adjust their behavior to match expectations for given social situations. Individuals act as information transmitters, simultaneously giving others intentional cues and giving off unintended signals. Individuals regulate their social behavior by becoming decision-making machines, these cues and signals serving as inputs and social mores functioning as "programming." This control process helps one to build and to manage a persona, to fashion an ever-changing public self tailored to the occasion.

CONTROL THROUGH COMMUNICATION

Alfred Chandler's work on the rise of corporate capitalism in the late nineteenth century has influenced an entire generation of business historians.[5] Between 1840 and 1920, Chandler argues, professional managers put in place modern business management techniques and made obsolete proprietary capitalists who prospered or not at the whim of external market forces. The business enterprise evolved from a creature of instinct forced to adapt to its economic environment to an organization capable of molding its economic surroundings.

As the firm grew in sophistication and complexity, so did its need for information and control. James Beniger has discovered a "crisis of control" in the mid-nineteenth century American economy, showing that the evolution Chandler describes was impossible without a corresponding late-nineteenth century "control revolution" in information processing and communication technology.[6] He is grandiose to claim, "We would have to go back at least to the emergence of the vertebrate brain if not to the first replicating molecule—marking the origin of life on earth—to find a leap in the capability to process information comparable to that of the Control Revolution." We should forgive this excess of enthusiasm, as Beniger draws much-needed attention to the ethos of control, establishing it as the hallmark of nineteenth and twentieth century business activity.

JoAnne Yates, in her study of American management between 1850 and 1920, draws upon Chandler and Beniger to examine how businesses established internal "control through communication." She examines several office technologies—the typewriter and vertical filing system, for example—to demonstrate how they undergirded "a new philosophy of management based on system and efficiency." This group of information technologies served as "a mechanism for managerial coordination and control of organizations."[7]

Chandler, Beniger, and Yates show that nineteenth and twentieth century business managers found communication technology—indeed, communication itself—indispensable to their control over external economic forces and internal organization. This was certainly true of the grain trade in the Midwest after 1850.

Grain traders employed the telegraph to obtain market prices over farflung areas. Speculators depended upon exclusive possession of such information to conduct their trades, and they almost always encoded it to ensure secrecy. For instance, a nine-word message from the 1850s, "bad, came, aft, keen, dark, ache, lain, fault, adopt," yielded information about several commodities:

> Flour market for common and fair brands of western is lower, with moderate demand for home trade and export. Sales 8000 barrels. Genesee at $5.12. Wheat prime in fair demand, market firm, common description dull, with a downward tendency; sales, 4000 bushels at $1.10. Corn, foreign news unsettled the market; no sales of importance made. The only sale made was 2500 bushels at 67 cents.[8]

Viewed as an information processor, the nineteenth century grain exchange was nothing more than a machine for setting prices. Its inputs were information streams giving various local supplies of and demands for grain. The trading floor processed these inputs, smoothing out local variations and yielding a uniform price as its output.

Such an information-processing machine was impossible without the telegraph; indeed, the exchanges were "the first business institutions which were actually created by electronic communication."[9] All of the major Great Lakes exchanges started trading with ten years of Morse's first telegraph message, within a few years after the telegraph reached their cities. The telegraph drastically reduced transaction and information costs, and the resulting economies of scale helped to centralize the grain trade within the exchanges. Futures trading performed a temporal as well as spatial centralization, reducing uncertainty and risk for transactions several months ahead. The outcome of all this was "market perfection," a market regulated by supply and demand, with little barrier to entry, and with prices known to all.[10]

This description is suitable for the grain trader and the economist, but it does not account for the hostility which hundreds of thousands of farmers directed toward the exchanges in the 1880s and 1890s.[11] The farmer looked toward the city with suspicion, venting his frustration upon the unseen grain broker who exercised parasitic control over his very livelihood. The telegraph made this control possible.

THE TELEGRAPH AND LANGUAGE

The telegraph made possible dramatic increases in information flow and yoked communication to new arenas of economic control.

These twin developments, however, have muted historical discussion of the telegraph's revolutionary impact on nineteenth century language. This lack is puzzling, since many contemporary Americans commented upon the telegraph's linguistic effects. Henry David Thoreau warned during its first decade of operation that the haste to construct a national telegraph system made it seem "as if the main object were to talk fast and not to talk sensibly."[12] The new communication device privileged speed over content because it was "the first electrical-engineering technology and, therefore, the first to focus on the central problem in modern engineering: the economy of a signal."[13] Brevity is the soul of wit, but it was also an economic imperative.

The telegraph performed three fundamental operations upon the language: compression, encryption, and standardization. Because every word cost money, the sender of a message worked to eliminate words of low information content. Modifiers, prepositions, and articles evaporated, leaving messages consisting of nouns, verbs, and numbers. Users of the telegraph eliminated rhetorical conventions found in written correspondence. Repetition, for emphasis or for clarity, became redundant and wasteful. Niceties of style and elegant turns of phrase were telegraphic extravagances. Formal greetings and closings wasted valuable transmission time. One observer writing in the *Democratic Review* forecast this trend in 1848, only four years after Morse's first message: "Now the *desideratum* of the Telegraph . . . is this—*How can the greatest amount of intelligence be communicated in the fewest words?*. . . Every useless ornament, every added grace which is not the very extreme of simplicity, is but a troublesome encumbrance." The author looked forward to the near future, when the written word will have matched the "Telegraphic style . . . , terse, condensed, expressive, sparing of expletives and utterly ignorant of synonyms."[14]

Americans between the Revolution and the Civil War fashioned a unique American language distinct from British English. This process of linguistic formation and differentiation guided them in their search for a national identity.[15] The commentator from the *Democratic Review* linked the telegraph to these twin efforts: "In these days of Yankee enterprise and activity, we want no prosaic Johnsonism; we can tolerate no dainty euphuist." Only writers who express themselves with "Yankee directness and concentration of . . . style" will have held the attention of the future republican reader. The author imagined the national character, mirrored through telegraphic language, as urban and Northern, enterprising and vigorous, masculine and youthful. This in opposition to effete "old gentlemen" sequestered "in the quiet solitude of their own libraries, undisturbed by the turmoil and bustle of the great world surging past their doors."[16]

The telegraph operated upon the language in another way, by treating it as a stream of symbols to be encoded. The use of Morse code to transmit messages is the obvious form of encryption. Morse recognized that some letters in English occur more frequently than others, and he used this statistical property to formulate his code. He assigned frequently occurring letters the shortest code symbols, while the infrequent letters received the longest code symbols. The telegraph was the first device to exploit the language's statistical properties in a systematic and intentional fashion.

Newspaper wire services throughout the nineteenth century employed a more important form of encryption, telegraph ciphers which telescoped lengthy stories into a few code words. Before the Civil War the British magazine *Once a Week* commented on the "remarkably elastic properties" of these ciphers. "A message of ten cipher words would expand to fifty or sixty, or even a hundred, when translated."[17] Correspondents reporting on Congressional proceedings used an imaginative and ironic cipher: the word "bacon" meant "a report was brought up from the Committee on Agriculture," and "bawl" stood for "an interesting debate followed, in which several honourable senators took part."[18] By connecting vivid and humorous images to the translated phrases, the cipher served as a mnemonic device which helped reporters to reconstruct stories quickly and accurately.

While this cipher conceptually linked many code words to their denoted English phrases, it sometimes introduced grave mistakes into final newspaper stories. The word "dead," for instance, meant "after some days' absence from indisposition, the honourable gentleman reappeared in his seat." This denotation provided reporters an easily remembered association between the cipher word, a senator's absence, and his figurative resurrection. This connection took on a life of its own when correspondents sent a story over the wire containing the phrase "John Davis dead." Many operators did not decode "dead" but read it literally. "Immediately the sad event was communicated all over the Union of the death of Davis, who, on the following day, had the privilege vouchsafed to but few persons, of learning what was the opinion of posterity upon his private life and public career."[19]

A similar mistake occurred toward the turn of the century during a period of diplomatic tension between the United States and Italy. A press association instructed its correspondents that the newly coined code word "dago" was to stand for "Baron Fava, the Italian Ambassador," providing reporters a memorable and jingoistic connection between the word and its referent. One operator, however, failed to receive this instruction and did not decode the word when it appeared in a dispatch. A prominent newspaper sparked a potential international incident the next morning when it reported the activities of "a dago," which "other equally reputable newspapers were at the same time attributing to the personal representative of the august sovereign of Italy."[20]

Equipment malfunctions and fallen lines introduced another sort of error into the final newspaper story. Such technical difficulties often blocked transmissions in the first few decades of telegraphy, and reporters became adept at fictional reconstructions of interrupted stories. One such breakdown in the 1850s stopped the cipher transmission of a Whig senator's speech concerning a bill under deliberation. Anxious to meet the deadline for the next morning's issue, the receiving reporters guessed how this senator might have reacted to this legislation. They relied upon their knowledge of congressional politics and their experience with this particular senator to concoct "a very violent speech against it." They guessed wrong, as "the honourable senator had spoken and voted in exactly the opposite direction."[21]

The telegraph, through wire-service journalism, encouraged a standardized and national language. James Carey has argued that the wire services required a "form of language stripped of the local, the regional and colloquial. . . . If the same story were to be understood in the same way from Maine to California, language had to be flattened out and standardized." The wire services replaced a colorful and local journalistic style using "the tall story, the hoax, much humor, irony and satire" with a distanced and dispassionate "language of strict denotation."[22] Historian Kenneth Cmiel demonstrates that "wild, undisciplined language" marked by "tall talk" and "fantastic overstatement in the service of self-aggrandizement" found its way into newspaper stories during the 1830s and 1840s. It did not jar their sensibilities when readers of the 1847 *New York Tribune* found out that a railroad stationmaster had "absquatulated with funds."[23] An observer at the turn of the century, however, noted that "the wires are . . . constantly shrinking the national vocabulary, hastening the retirement of words of the less useful sort."[24] Apparently "absquatulated" was one.

As newspapers increasingly relied on the wire services for distant news, sensational reporting and rhetorical flourish disappeared from national stories. Several Pittsburgh newspapers employed correspondents to cover a notorious murder trial in Cambridge, Massachusetts, in 1850. The papers instructed them to mail full reports of the daily courtroom drama for publication, but urged them to telegraph the verdict as quickly as possible. Eventually the wire services standardized the format of Washington and New York dispatches, and stories based on them appeared identically in many Western and Southern newspapers.[25]

The telegraph also restricted forms of punctuation and all but eliminated sentences of complex construction, anticipating the style of twentieth century writers such as Hemingway and Vonnegut. Robert Lincoln O'Brien noted in 1904 that a sentence beginning with an adverbial phrase was ambiguous over the wires, since the phrase might have also pertained to the previous sentence. A sentence beginning with a subject noun did not suffer this ambiguity. In telegraphic transmission "only the most obvious stops can be depended upon," forcing writers to use sentences which "punctuate themselves" by starting "directly, and even abruptly . . . Much of the graceful elision of one sentence into the next" withered over the wire, leaving only functional and plain prose. "Where each sentence stands out as direct as a brick the literary passage will have the aspect of a brick wall."[26]

Mr. O'Brien, writing in 1904, clashed sharply with the anonymous observer who wrote in the 1848 *Democratic Review*. The mid-century writer had eagerly anticipated a condensed and vigorous writing style, while O'Brien mourned the demise of elegant prose. Their disagreement mirrored deeper changes in how nineteenth century Americans conceived of and deployed the national language. Antebellum Americans sought to construct a national idiom and a distinctive literature partly to obscure or to eradicate sectional and class differences. Whitman's rugged and expansive verse aptly reflected this desire for an all-inclusive national identity. At the turn of the century, language and literature no longer served these purposes. The Civil War, Reconstruction, and overseas imperialism all but eliminated the drive to gloss over sectionalism, while flashpoint class tensions and unprecedented Southern and Eastern European immigration called into question the very notion of a national language and character. Mr. O'Brien's discomfort with telegraphic style resonated with his educated, middle-class audience because they used literary taste as a distinguishing mark in their construction of a class identity.[27]

THE MATHEMATICAL ABSTRACTION OF THE COMMUNICATION ACT

While many nineteenth century Americans debated the telegraph's effect upon the language and literary style, others developed ideas which depersonalized the notion of communication itself. Telegraph electricians and telephone engineers constructed these notions precisely because the technological and economic imperatives of electrical communication encouraged them to conceive of the message as an arbitrary waveform traversing an abstract medium, a mental feat impossible with older forms of communication. In fact, this forms the conceptual foundation upon which several devices used in telegraphy and telephony rest.

Multiple telegraph systems comprised the major communication innovation between the invention of the telegraph in 1837 and the telephone in 1876.[28] Multiple telegraphy allowed a single telegraph wire to transmit more than one message at a time. The duplex system introduced in 1872 provided the means to transmit two messages over the same wire in opposite directions, and Edison's 1874 quadruplex system enabled the transmission of four messages, two in each direction. Both systems operated on a system of timesharing in which a mechanical device switched the line between the two or four telegraph stations several times a second. The acoustic telegraph, forerunner to the telephone, operated in a different fashion. It generated oscillations which excited the wire at several wavelengths, each carrying one telegraph message. One wire was able to transmit dozens of simultaneous messages. Acoustic telegraphy, in fact, provided the basis for Elisha Gray's and Alexander Graham Bell's pioneering work on the telephone; both realized that these wavelengths when combined were capable of transmitting the human voice.

These systems gave telegraph companies the ability to increase message traffic without installing more wires or equipment. Western Union reported in 1883 that 25 percent of their total mileage was "phantom line" created by duplex and quadruplex systems. By increasing a line's capacity twofold or fourfold, multiple telegraph systems reduced operators' idle time; seen from the telegrapher's perspective, then, these systems were a work speedup. Multiple telegraph systems formed an important part of the "new economic order coalescing in the era" which moved toward "high-volume flow, centralization, and thoroughgoing rationalization" as well as toward Taylorization of the workforce.[29]

Behind these economic trends, however, lay another meaning, a higher level of abstraction in thinking about the communication process. Before the introduction of the duplex and quadruplex, one telegraph line carried only one message; there still existed a somewhat comprehensible link between the message and the medium. The inventors of these systems broke this link. They conceived of the two or four messages on the wire as a set of time series subject to mechanical manipulation, and they visualized the telegraph line as a medium to be apportioned among several messages. Another conceptual refinement occurred with the introduction of acoustic telegraphy. Instead of apportioning wire time among several messages, it assigned each message a unique wavelength. It was similar in principle to the allocation of the radio spectrum to twentieth century radio broadcasting. It also bore a striking resemblance to Fourier analysis, the branch of mathematics which models processes of this sort. Fourier had formulated his theory in the early nineteenth century, yet it is nearly certain that telegraph inventors of the 1870s were unaware of it. It remained in the domain of theoretical physics until turn-of-the-century engineers drew upon it to advance their understanding of alternating-current generating systems and wireless telegraphy.

This conceptual convergence is striking because it reveals the extent to which telegraph electricians abstracted the communication process. The capacity for such abstraction is common to twentieth century communication engineers, trained as they are in mathematics and physics. It is somewhat surprising to find that telegraph electricians of the 1870s, rooted in shop practice, embodied such abstractions in their mechanical switching devices.

Engineers who worked on long-distance telephony in the first decade of this century form for us a bridge between these telegraph electricians and their mid-20th century counterparts. The electricians imparted a practical and experiential emphasis to early electrical engineering, but university-trained engineers in the 1880s and 1890s transformed it into an engineering science with secure foundations in mathematics and physics. By the turn of the century electrical engineers defined themselves as professionals who systematically reduced rigorous theoretical principles to practical applications.

Columbia University physicist Michael Pupin and AT&T engineer George Campbell independently invented the loading coil which made possible transcontinental telephone conversations. Both drew upon theoretical physics to surmount the long-distance barrier confining telephone transmission to 1,200 miles. In 1900 the Patent Office awarded Pupin a patent for the device, and AT&T purchased his rights later that year to avoid legal disputes over priority. Their engineers immediately installed the devices on a long-distance line between Bedford, Massachusetts, and Brushton, Pennsylvania, and they perfected the technology over the next several years.[30] By increasing the clarity and volume of the human voice, the loading coil helped AT&T to steal away Western Union's dominance in long-distance communication. The coil and related innovations ensured the obsolence of telegraphy by 1920.

Prior to the loading coil, few innovations in communication technology drew their conception from mathematical physics. After it, few did not. Its inventors reduced to practice a set of equations which described transmission as a combination of the signal's specific waveform and the telephone line's electrical properties. Their work represented a further abstraction in how engineers conceived of the communication process. This theoretical apparatus gave turn-of-the-century engineers the ability to modify the properties of the transmission line, a capacity which telegraph electricians of the previous quarter-century had not shared.

Technical men in the late nineteenth and early twentieth centuries gained increasingly sophisticated control over the communication process, first through practical experience, then through the rigorous application of scientific theory. They struck a hollow bargain, achieving this only through an abstract process which removed the human element from discussion. These abstractions formed the conceptual foundations of cybernetics and information theory.

THE HUMAN AS COMPONENT: CYBERNETICS AND INFORMATION THEORY

American electrical engineers during World War Two contributed to the war effort on two fronts. They designed fire-control devices such as radar and the proximity fuse, and they standardized and automated heavy industry upon which war production rested. In both sets of problems they employed the ethos of control through communication to augment or to replace human skill and judgment. Fire-control devices which automatically tracked targets, aimed weapons, and detonated shells foreshadowed postwar efforts toward machine cognition.[31] Wartime automation evolved into postwar work on electronic machine-tool control and robotics.[32]

Warren Weaver, one of the founders of postwar communication

and control engineering, identified three levels to the communication process:

1. LEVEL A. How accurately can the symbols of communication be transmitted? (The technical problem.)
2. LEVEL B. How precisely do the transmitted symbols convey the desired meaning? (The semantic problem.)
2. LEVEL C. How effectively does the received meaning affect conduct in the desired way? (The effectiveness problem.)[33]

Weaver's hierarchy crystallized the process of abstraction begun in the 19th century and required a key conceptual prerequisite: the irrelevance of the human elements of communication—the message content and the communicators. He made it clear that engineers viewed communication as the efficient transmission of a compressed symbol stream for purposes of control.

While Weaver's theory dealt exclusively with "the technical problem" of accurate signal transmission, Norbert Wiener expanded his work and that of other communication engineers to found cybernetics, a science which reduced human physiology and psychology to machinelike characteristics. To Wiener a computer logic circuit was "almost an ideal model of . . . the nervous system," and the brain was "nothing but a mechanism . . . hav[ing] its precise analogue in the computing machine." Human memory "has its parallel in . . . artificial memories for the machine."[34] Wiener carried this logic to its ultimate conclusion: in a passage worthy of Mary Shelley he gave a blueprint for the construction of a cybernetic organism—cyborg:

> It has long been clear to me that the modern ultra-rapid computing machine was in principle an ideal central nervous system to an apparatus for automatic control; and that it input and output . . . might very well be . . . the readings of artificial sense organs, such as photoelectric cells or thermometers, and the performance of motors and solenoids. With the aid of strain gages or similar agencies to read the performance of these motor organs and to report, to "feed back," to the central control system as an artificial kinesthetic sense, we are already in a position to construct artificial machines of almost any degree of elaborateness of performance.[35]

Wiener suspected that his theory, when taken to its extreme, reduced human qualities to engineering parameters. He explicitly wished to limit cybernetics to productive and humane endeavors such as the construction of prosethetic limbs and the cure of neurological diseases. His work contained a tension between the science's humane uses and its destructive potential, between the moral need to limit its applications and the impossibility of doing so. Cybernetics contained "great possibilities for good and for evil. We can only hand it over into the world that exists about us, and this is the world of Belsen and Hiroshima."[36]

Information theorists blurred the boundary between human and machine from the other side, by imparting human qualities to computing machines. The 1962 annual meeting of the Institute of Radio Engineers focused on information theory and the construction of automata. One optimistic presenter, computer scientist J. Presper Eckert, believed that engineers stood on the brink of constructing "a self-reproducing automaton which can improve itself." Thanks to previous researchers, "memory, eyes, ears, hands, and logic" were "about as good" in automata as in humans. However, computer scientists still needed to improve their "recognition ability, certain types of information retrieval, and the ability to taste and smell." R. M. Page of the Naval Research Laboratory concluded that the "mechanization" of "the functions of the human brain" was technically possible, if not inevitable.[37]

Information theorists attempted to mechanize two functions of the human brain: the use of language and the ability to learn. Early computer scientists studied linguistics to construct machines capable of carrying out programmed commands. Nearly all researchers felt that natural languages such as English were too imprecise and redundant for this purpose, so many examined their structure to improve their usefulness as information-processing languages. Harvard University engineer George Miller conducted research into the reconstruction of distorted English texts and discovered that "the average person . . . will not be able to correct passages perfectly if more than 10% of the characters are mutilated" in a random fashion. He concluded that "these results correspond to a lower bound of 60% for the redundancy of printed English."[38] He and a research team also examined the statistics of written English passages; they counted the "number of times each different word occurs in a long passage," listed them "in order of decreasing frequency," and gave each "its rank in that order." They verified that, "if r is the rank and f is the frequency of occurrence, then $rf=k$, where k is a positive constant depending upon the size of the sample." They combined this rule with "a two-state Markov process that will generate strings of letters interrupted by spaces" and obtained a passable imitation of English text.[39] In 1976 Yale engineering professor William Bennett calculated that it would take over a trillion years for a trillion proverbial monkeys, randomly typing ten keys per second, to generate Hamlet's famous line, "To be, or not to be: that is the question." However, when computer programming constrained a virtual monkey to employ the built-in redundancies of English and the statistical structure of Hemingway's *A Farewell to Arms,* it immediately typed an obscene proposal.[40]

Other researchers worked toward the construction of machines with decision-making abilities. Psychologist Donald Campbell claimed that, "For the time being, perhaps human communication units have a monopoly on this capacity, although it would be possible to construct a machine for this purpose."[41] Campbell undertook a research program to remove human decision-making from communication links because humans made mistakes "over and above random imperfection of performance."[42] Individuals, in other words, did not follow directions as well as machines. He compared learning in humans and machines" not to deny that man is a machine, but rather to emphasize what *kind* of machine man is, and how he differs from certain mechanisms that might be substituted for him in the communication network." He found that "early in learning the task the human performance was like that of a one-integrator system with a feed-forward loop," while in later stages "the human performed analogously to a two-integrator system." Unfortunately, "as certain machines are made more complexly adaptive to their purposes, their biases become more human."[43]

CONCLUSION: COMMUNICATION AND SELF-IDENTITY

Neither the abstraction of communication nor the blurring of human and machine remained confined to the world of engineers and computer scientists. Both molded how Americans conceived of the communication process and of themselves. By way of conclusion this article provides two brief examples: ways in which the ideology of control through communication shaped the first fifty years of telephone use, and how notions borrowed from communication engineers affected self-identity in the mid-twentieth century.

Businessmen formed the major market for telephone subscription until the First World War. Since the telegraph was the first form of electrical communication, early subscribers and salesmen both conceived of the telephone as a sort of speaking telegraph with its attendant pragmatic uses. The industry marketed it as "a practical device for attaining practical ends," and defined its proper role as "the conduct of business or related activities" among urban managers.[44] Telephone companies also sold the instrument to a secondary market, the middle- and upper-class housewife, anticipating her to use it as a household management tool. A 1910 Bell advertising campaign to boost residential subscriptions dwelled only on its practical uses around the home: telephoning retailers, ordering groceries, and extending brief social invitations.[45]

Early residential subscribers paid a flat monthly fee for telephone service, regardless of the number of calls made or their length. The assumption that calls were brief exchanges of information, not extended conversations, undergirded this rate structure. Many customers did not conform to this expectation. As early as 1881 a telephone executive lamented that the flat rate allowed residents "to use the wires as they pleased without incurring additional expense," especially to conduct "large numbers of communications of the most trivial character."[46] To discourage such frivolity many telephone companies before the turn of the century placed official time limits of five minutes on calls.[47] Such efforts proved futile, and by the 1920s most local companies began to charge subscribers per call. The telephone industry, in modifying its rate structure, acquiesced to uses they had previously discouraged.

Industry executives identified women as the source of frivolous and unnecessary telephone visiting, and their campaigns to curtail such uses nearly always labeled the caller as "she." When the Bell System later promoted the telephone's sociable uses in the 1920s and 1930s, their advertisements largely targeted women.[48] Carolyn Marvin claims that men used the instrument to relay "brief, efficient transmissions" of "instrumental information," while women prized "redundant, frivolous, playful, and abundant" conversations which cemented "the personal relationship that was the real subject of any electrical conversation."[49]

The gap between the telephone's twin uses—functional and sociable—opened up an arena for new social roles and behaviors, forming a terrain in which "innovative experiments could take place in all social relations."[50] Communication over the telephone eliminated face-to-face visual cues which marked class and racial affiliation, striking at the heart of a social order in which individuals used them to organize personal relationships and to regulate social distance. Turn-of-the-century confidence men, for instance, used established forms of communication such as letters and personal visits to gain the victim's trust, using these contacts to establish himself as a social superior to the victim. The criminal then employed the telephone to commit the fraud.[51]

The telephone remapped the social world by changing how individuals oriented themselves in relation to others. In an analogous fashion, values embedded in electrical communication modified conceptions of self-identity and the public presentation of self in the twentieth century. In the late 1950s sociologist Erving Goffman borrowed concepts from communication engineering to explain the dynamics of social interaction. He started with the insight that an individual in a social encounter simultaneously gives intentional cues and gives off unintended signals. Both kinds of information transmit the individual's expectations of the gathering and project a public persona to others. On the conscious level an individual in public tailors "his activity so that it will convey an impression to others which it is in his interests to convey." One fashions this impression "to control the conduct of others, especially their responsive treatment of him."[52] Unconscious signals, such as attentiveness and posture, reveal to the audience the truth or falsity of one's self-portrait. In Goffman's dramaturgical world the individual becomes a performer, "a harried fabricator of impressions," playing a character possessing "sterling qualities" which "the performance was designed to evoke."[53]

Audience members react to the impression that one seeks to convey. These feedback signals, the audience's unintended response to this impression and intentional "lines of action they initiate to him,"[54] help one to adapt to the given social interaction. Public behavior, or "the arts of impression management,"[55] is a perpetual feedback-and-control loop, "a kind of information game—a potentially infinite cycle of concealment, discovery, false revelation, and rediscovery."[56]

Goffman used a communication-and-feedback model to describe social behavior and public identity in late twentieth century America. Historian Warren Susman has traced this process from the turn of the century onward, arguing that our vision of the ideal social self has changed. Americans in the nineteenth century valued "character," or moral stature, but Americans early in this century prized "personality," or the ability to be well-liked.[57] Character "provided a method of presenting the self to society, offering a standard of conduct that assured interrelationship between the 'social' and the 'moral.'"[58] This definition of worthy conduct, rooted in self-denial and obedience to moral strictures, regulated public behavior in the producer society of the previous century. Susman claims that the transition to a consumer society weakened this link, and he dates the first signs of this to the first decade of this century. By the 1920s the individual possessing personality—appearance, manners, poise, and charm—formed the ideal social type. Self-fulfillment and popularity became the standards by which one measured happiness in twentieth century consumer society.

Susman correctly asserts that "both visions of self . . . shaped the very nature of the culture" and suited "the needs of a particular social structure."[59] Both notions of self-worth also reflected the means available to the individual to fashion a public self, the ways in which one communicated an identity to others. In the relatively stable producer society of the nineteenth century, character—marked by "citizenship, duty, democracy, work, building, golden deeds, outdoor life, conquest, honor, reputation, morals, manners, integrity, and above all, manhood"[60]—formed the basis for one's public stature. In the fast-paced consumer society of the twentieth century, personality—described by the adjectives "fascinating, stunning, attractive, magnetic, glowing, masterful, creative, dominant, forceful"[61]—measured one's standing among others. Social success for one possessing "personality" depended upon one's ability to obtain information about others so as to regulate behavior, to give others selected information so as to convey a favorable impression of oneself, and to gain a measure of control over the given social situation. These abilities bear striking resemblance to many of the values implicit in communication engineering. This is not to suggest that innovations in electrical communication caused this shift in self-identity, but it does call for an examination of their correlation.

Goffman concluded his work with a description of social interaction which overturned the idea of stable and autonomous selfhood:

> In analyzing the self then we are drawn from its possessor, from the person who will profit or lose most by it, for he and his body merely provide the peg on which something of collaborative manufacture will be hung for a time. And the means for producing and maintaining selves do not reside inside the peg; in fact these means are often bolted down in social establishments . . . The self is a product of all these arrangements, and in all of its parts bears the marks of this genesis.[62]

If we are indeed these contingent and socially constituted selves, we are obligated, with whatever agency may be left to us, to understand how and why this transformation occurred. This inquiry must include a critical examination of the technologies we have employed to communicate with each other—why we build them and how we use them.

NOTES

The author wishes to thank Paul Israel for valuable assistance with sources on telegraphy and language; and Catherine Kelly, Carroll Pursell, Jonathan Sadowsky, and Jeffrey Yost for their thoughtful comments on earlier drafts of this article.

1. I am borrowing this phrase from JoAnne Yates. *Control Through Communication: The Rise of System in American Management,* Baltimore, Johns Hopkins University Press, 1989.

2. Norbert Wiener, *Cybernetics: Or Control and Communication in the Animal and the Machine,* New York, John Wiley and Sons, 2nd edition, 1961, p. 11. MIT published the first edition in 1948.

3. Today these disciplines, loosely defined, have changed names. Cybernetic concepts are fundamental to robotics and biomedical engineering, while information theory forms the basis for computer "artificial intelligence" research.

4. See especially his works *The Presentation of Self in Everyday Life,* Garden City, Doubleday Anchor Books, 1959; and *Behavior in Public Places,* New York, The Free Press of Glencoe, 1963.

5. Alfred Chandler, *The Visible Hand: The Managerial Revolution in American Business,* Cambridge, Belknap Press of Harvard University, 1977.

6. James R. Beniger, *The Control Revolution: Technological and Economic Origins of the Information Society,* Cambridge, Harvard University Press, 1986, pp. 35–6.

7. JoAnne Yates, *Control Through Communication: The Rise of System in American Management,* Baltimore, Johns Hopkins University Press, 1989, pp. xv–xx.

8. E. McDermott, "Telegraph Reporting in Canada and United States," *Once a Week* 3, Sept. 1, 1860, pp. 258–9.

9. Richard B. Du Boff, "The Telegraph in Nineteenth-Century America: Technology and Monopoly," *Comparative Studies in Society and History* 26, October 1984, p. 574.

10. For insight into the operations of the grain trade, see Cedric B. Cowing, *Populists, Plungers, and Progressives: A Social History of Stock and Commodity Speculation, 1890–1936,* Princeton, Princeton University Press, 1965. For a description of Chicago's economic hegemony over the Midwest and Great Plains, see William Cronon, *Nature's Metropolis: Chicago and the Great West,* New York, W. W. Norton and Company, 1991.

11. For a contemporary fictional account, see Frank Norris's short story "A Deal in Wheat," in which low wheat prices on the Chicago

Board of Trade force a Kansas farmer to default on his mortgage. The farmer moves his family to Chicago, becomes destitute, and is unable to buy bread because speculators attempting to run a corner have driven up the price of wheat. William Cronon describes how the Chicago Board of Trade helped the city attain economic hegemony over much of the Great Plains; *Nature's Metropolis: Chicago and the Great West,* New York, W. W. Norton and Company, 1991.

12. Henry David Thoreau, *Walden,* Boston, Houghton-Mifflin, 1957, p. 38.

13. James W. Carey, "Technology and Ideology: The Case of the Telegraph," *Prospects: The Annual of American Cultural Studies* 8, p. 304.

14. Italics in the original. "Influence of the Telegraph Upon Literature," *Democratic Review* 22, May 1848, pp. 411, 412.

15. Kenneth Cmiel, "'A Broad Fluid Language of Democracy': Discovering the American Idiom," *Journal of American History* 79, Dec. 1992.

16. Influence of the Telegraph Upon Literature," pp. 412–3.

17. E. McDermott, "Telegraph Reporting in Canada and United States," *Once a Week* 3, Sept. 1, 1860, p. 258.

18. Ibid., p. 259.

19. Ibid.

20. Robert Lincoln O'Brien, "Machinery and English Style," *Atlantic Monthly* 94, October 1904, p. 466.

21. McDermott, p. 259.

22. Carey, p. 310.

23. Cmiel, p. 920.

24. O'Brien, p. 468.

25. Richard B. Kielbowicz, "News Gathering by Mail in the Age of the Telegraph: Adapting to a New Technology," *Technology and Culture* 28, January 1987, p. 35.

26. Ibid., pp. 468–70.

27. For an interesting and iconoclastic contemporary view, see Thorstein Veblen's essay "The Higher Learning as an Expression of Pecuniary Culture," in *The Theory of the Leisure Class,* originally published by the MacMillan Company in 1899. Veblen argued that the acquisition and use of "elegant diction" indicated the "industrial exemption" of the leisure class, while the productive classes necessarily employed "direct and forcible speech." (New American Library edition, pp. 256–8.)

28. Paul Israel provides a clear discussion of the technical characteristics, business decisions, and legal struggles surrounding mul-

tiplexing. Edwin Gabler has written a nuanced labor history of telegraph operators; he gives an interesting account of the effects of multiplexing upon their working conditions and attitudes. Paul Israel, *From Machine Shop to Industrial Laboratory: Telegraphy and the Changing Context of American Invention, 1830–1920,* Baltimore, Johns Hopkins University Press, 1992, especially pp. 134–141, p. 113, and p. 141. Edwin Gabler, *The American Telegrapher: A Social History, 1860–1900,* New Brunswick, Rutgers University Press, 1988, especially pp. 53–4.

29. Gabler, pp. 53–4.

30. I base much of my discussion of the loading coil on Neil Wasserman's book *From Invention to Innovation: Long-Distance Telephone Transmission at the Turn of the Century,* Baltimore, Johns Hopkins University Press, 1985. Wasserman combines an understanding of the technology with a clear explanation of the business decisions taken by AT&T to deploy the new device.

31. See A. Michal MacMahon, *The Making of a Profession: A Century of Electrical Engineering in America,* New York, The Institute of Electrical and Electronic Engineers, 1984, pp. 195–206.

32. David Noble provides a provocative account of wartime and postwar work on factory automation in *Forces of Production: A Social History of Industrial Automation,* New York, Alfred E. Knopf, 1984. He argues that engineers and industrialists collaborated to deskill machinists through electronic machine control.

33. Warren Weaver, "Recent Contributions to the Mathematical Theory of Communication," in Claude Shannon and Warren Weaver, *The Mathematical Theory of Communication,* Urbana, University of Illinois Press, 1964, p. 4. This book contains two sections, Weaver's nonmathematical discussion of communication engineering, and Shannon's revision of his seminal paper published in the late 1940s in the *Bell Labs Technical Journal.*

34. Norbert Wiener, *Cybernetics: Or Control and Communication in the Animal and the Machine,* New York, John Wiley & Sons, 2nd edition, 1961, p. 14. MIT published the first edition in 1948.

35. Ibid., pp. 26–7.

36. Ibid., p. 28.

37. Quoted in MacMahon, p. 246.

38. George A. Miller and Elizabeth A. Friedman, "The Reconstruction of Mutilated English Texts," *Information and Control* 1, 1957, p. 38.

39. George A. Miller, et al. "Length-Frequency Statistics for Written English," *Information and Control* 1, 1958, pp. 370, 372.

40. Jeremy Campbell, *Grammatical Man: Information, Entropy, Language, and Life,* New York, Simon and Schuster, 1982, pp. 116–7.

41. Donald T. Cambell, "Systematic Error on the Part of Human Links in Communication Systems," *Information and Control* 1, 1958, p. 347.

42. Ibid., p. 334.

43. Ibid., p. 339.

44. Claude Fischer, *America Calling: A Social History of the Telephone to 1940,* Berkeley, University of California Press, 1992, p. 69.

45. Claude Fischer, " 'Touch Someone': The Telephone Industry Discovers Sociability," *Technology and Culture* 29, January 1988, p. 39.

46. Quoted in Fischer, "Touch Someone," p. 48.

47. Ibid., p. 48.

48. Ibid., p. 51–2.

49. Carolyn Marvin, *When Old Technologies Were New: Thinking About Electric Communication in the Late Nineteenth Century,* New York, Oxford University Press, 1988, pp. 24–5.

50. Ibid., p. 108.

51. Ibid., pp. 92–5.

52. Goffman, *The Presentation of Self in Everyday Life,* pp. 3–4.

53. Ibid., p. 252.

54. Ibid., p. 9

55. Ibid. This is the title for Chapter VI, pp. 208–37.

56. Ibid., p. 8.

57. Warren Susman, " 'Personality' and the Making of Twentieth-Century Culture," in *Culture as History: The Transformation of American Society in the Twentieth Century,* New York, Pantheon Books, 1984.

58. Ibid., p. 273.

59. Ibid., p. 280.

60. Ibid., pp. 273–4.

61. Ibid., p. 277.

62. Goffman, *The Presentation of Self in Everyday Life,* p. 253.

7

MEDICAL CYBORGS: ARTIFICIAL ORGANS AND THE QUEST FOR THE POSTHUMAN
Chris Hables Gray

[W]e can predict that the number of artificial organs will increase exponentially and that the social and economic consequences will touch the lives of nearly all citizens of the developed world.—Dr. Willem J. Kolff (1979b, p. 8)

INTRODUCTION: CYBORG SOCIETY AND SOCIETIES OF CYBORGS

We can agree with Dr. Kolff that the number, and importance, of artificial organs will increase exponentially, but just what are some of the consequences of this proliferating technology? History can't give a definitive answer but it can offer some potentially valuable insights into the possibilities. That is the goal of this chapter, to find clues about our possible futures in the details of the recent past.[1]

The second half of the twentieth century has seen an incredible proliferation in the creation and use of artificial implants and organs in medicine, especially in Europe, North America, and Japan. Many people are familiar with artificial limbs, false teeth, cardiac pacemakers and the still unperfected artificial heart, but the range of biomedical prosthesis is actually much greater then most people realize. Inert attachments and aids, such as crutches, canes, wooden legs, hooks for hands, and false dentures of various material (wood, metal, bone, and animal teeth), have existed for thousands of years, but active attachments or implants that interrelate cybernetically (exchanging information and sometimes also energy) with human bodies, making them cybernetic organisms or cyborgs,[2] are of much more recent vintage.

The importance of cyborgs in contemporary society is not widely recognized despite the work of such pathfinders as Donna Haraway (1985, 1990a) and Levidow and Robbins (1989). Robots and disembodied artificial intelligences are deemed much more interesting theoretically by many observers, despite the fact that their realization is actually quite distant, while in most respects many of us are already cyborgs. The crux of it is that cyborgism represents a new type of human-machine relationship. Here an crucial distinction should be made[3] between *cyborg society* and a *society of cyborgs*. Cyborg society refers to the intimate and pervasive merging of humans and machines throughout high technology culture, a development that is central to the theories of Donna Haraway. The growing recognition that we live today in a machine-human society has even led to the development of a "cyborg anthropology" (Dumit et al. 1992). At the same time, many people, especially those who live in these cyborg societies, are actual cyborgs themselves.

For this article I will focus only on cyborg technologies that fit a more rigorous definition of cyborgism. Many developments that reflect the human attempt to remake the body but that involve human-machine interactions that are less than the full cybernetic integration that defines pure cyborgs[4] will not be dealt with here. These include interventions such as: dieting ($33 billion spent in the United States in 1990); weight training ($1 billion worth of exercise machinery sold each year in the United States); autografts for transsexual and other reconstructive surgery; or even cosmetic surgery, which 1.5 million people in the United States submit to each year (including 100,000 nose jobs, 250,000 liposuctions, 75,000 face lifts, and 130,000 breast implants for a total of over 2 million!) (Kimbrell 1992, p. 53). I also won't discuss the current programs to perfect test-tube babies (artificial wombs), nor genetic manipulations, nor the continuing widespread use of immunizations and other biochemical interventions, even though in almost all respects they do fulfill the definition of a cyborg and they are certainly part of our cyborg society.

While some of the technologies developed for disabled people will be discussed, the full range of computer and mechanical interfaces in this area will also not be described. Nor will I detail other centers of cyborg production in our culture, such as the mass media or the

Medical Cyborgs

military. I have written about the genealogy of cyborgs in fantastic literature and military projects already (Gray 1989, 1993). This article deals only with contemporary medical cyborgs and their antecedents, especially prosthetic implantations.

A quick tour of the human body, from the head to the toes, will offer some idea of how many ways medicine can turn humans into cyborgs. In the head the most advanced artificial implants are platinum or glass electrodes used as cochlea replacements (Gerhardt and Wagner 1987). They pick up sound waves and relay them directly to the auditory nerve. Interfacing this sound information with implanted computers for speech processing is also being developed (Stiglebrunner et al. 1987). Neural probes are often installed semipermanently in trauma victims to monitor the pressure of cerebral fluids and for the direct administration of medications. There is also significant research on a so-called "brain pacemaker" to treat neurological dysfunctions with timed, or evoked, electrical and biochemical stimulations (White 1992).

Artificial joint implants in the United States alone in 1985 were 110,000 hips, 65,000 knees, and 50,000 others (Nosé 1985, p. 7). There is extensive use of artificial veins, arteries, and bones as well.

Implanted electrical stimulation systems have a very broad range of applications and they are growing in sophistication (Mayer et al. 1987). As of 1990 (Nosé) they were being used to stimulate the diaphragms of children and adults to combat sleep and breathing disorders, to induce micturition, improve defecation, facilitate penile erection in paraplegics, and to stimulate atrophied muscles in paraplegics and quadriplegics. When used with exoskeleton legs, gait sensors, and computer walking programs they can sometimes even allow paraplegics to walk. Treating facial and other paralysis with nerves, muscles and electromechanical stress gauges implanted together is being studied (Broniatowski et al. 1991).

Complex biomaterials are being developed (especially polymers) that mimic biological tissues such as glands, cells, skin, bones, and cartilage. They are often "hybrids" that grow transfectec mammalian cells into polymeric gels or scaffolds (Hoffman 1991). Since 1969, when charcoal granules were put in ultrathin membranes and then injected into the blood stream for blood detoxification, there has been an increasing development of artificial cells including prototypes containing insulin, antibodies, anti-cholesterol compounds, enzymes, and urea converters. Research on artificial red blood cells and full artificial blood continues as well (Chang 1991). To understand the possible implications of this wave of cyborg medicine a few specific technologies will be considered: artificial limbs and genitalia, the artificial kidney, the artificial liver, and the artificial heart. Each case reveals important specifics about cyborgs of today and implications about cyborgs of tomorrow.

ARTIFICIAL LIMBS AND GENITALIA—THE SEARCH FOR WHOLENESS

The human being and his assistive device comprise a man-machine system.
—(Reswick and Vodovnik 1967, p. 5)

The ideal of physical perfection has always required completeness and symmetry; a missing body part renders a person unusual, vulnerable, even ugly. Consequently, since antiquity, amputees have sought to correct this unacceptable condition.
—(Romm 1988, p. 158)

The first record of an artificial limb is the Greek myth about the murder of Zeus' grandson, Pelops. Tantalus, curious about the ability of the gods to differentiate between animals and humans, served his flesh to the assembled Olympians as an experiment. Demeter ate part of Pelops' shoulder. When she discovered the accidental cannibalism she restored Pelops to life and replaced his shoulder with an ivory one (Romm 1988, p. 158). The earliest full account of an actual prosthesis is also from Greece. Herodotus (484 B.C.) reported that a man named Egesistratus amputated his own foot to escape imprisonment and that he later replaced it with an artificial limb (Putti 1929, p. 112). There are other similar accounts, and illustrations on pottery or walls, of amputees using prosthesis and crutches from throughout the ancient Mediterranean world. But it wasn't until the fifteenth and sixteenth centuries that more complex artificial limbs were developed. Around this time a number of inert and working hands were built in Italy, Turkey, Germany, and France. A master smith known as "le petit Lorrain" built one of the most complex examples. It allowed for flexing all four fingers. He also made an upper arm that bent at the elbow (Putti 1929, pp. 112–3). Artificial legs were also improved during this period and there is even one example of the first known prosthesis designed for a congenital defect instead of the traditional artificial limbs for victims of traumatic wounds from war and work.

Only recently has there been significant improvement over the prosthesis of the Middle Ages. In 1967, one leading researcher admitted that

> [I]t seems to me that significant progress in externally powered limbs will be made only when it becomes possible to link the central nervous system 'on line' with the prosthetic control system. (McKenzie 1967, p. 3)

Even in the late 1960s, Dr. D. S. McKenzie, the director of the British Ministry of Health's Biomechanical Research and Development Unit, had to admit that "a well-designed hook is more functional than any of the many so-called functional hands" and that it was "very doubtful indeed" that self-powered cybernetic limbs were any better than "conventional body-powered prosthesis" (1967, pp. 1–2). However, he foresaw that work on position-controlled servos and pressure-demand pneumatic valves could be regarded "as among the first breaches in the man-machine interface" (p. 3). He was right. By the 1990s not only had powered prosthesis become much more effective, but in many ways the human nervous system had indeed been brought "on line" through electrical stimulation and monitoring that allowed amputees, paraplegics, and even quadriplegics to control limbs through either muscle or nerve activity or both.

Sensors on and in muscles can now (since 1966) pick up electromyographic (EMG) signals and convert them to specific commands for the powered prosthesis (Reswick and Vodovnik 1967, p. 7; Graupe and Kohn 1994). While direct electroencephalographic (EEG) signals from the brain are not as easily controlled, nor understood, they are also being used experimentally, especially by the military (Gray 1989). As mentioned earlier, this work has advanced to the point where some disabled people can actually walk with the help of exoskeletons and where numerous important physiologic activities can be encouraged, and at times even controlled, through electrical stimulation.

Less invasive are projects to use head motions (tracked by accelerometers) and voice commands to initiate preprogrammed actions that control computer-directed prosthesis and slave robot arms for the disabled (ACM 1992) or for military pilots and drivers (Gray 1989). The disabled can call up eating motions, telephone dialing, and any number of domestic tasks. The military has focused on targeting, launching weapons, flying and driving, visual displays, and damage control activities.

As medical technology and practice improved in the late nineteenth century there were a growing number of attempts to treat malfunctions of that particularly male limb, the penis. The history of treatments for impotence probably goes back as far as the history of amputations and crutches, but early attempts at curing this male malady were always limited to the use of internal medicines (potions and herbs actually) and the use of surrogate partners (preferably virgins, at least according to the Bible where King David's unsuccessful treatment is described). In 1688, Regneri De Graaf discovered that he could induce erections in the dead by injecting the penis with fluid from the forerunner of the modern syringe, which he invented (Gee 1975).

Little applicable progress was made until the 1900s when various doctors began direct surgery on the penis. Between 1908 and 1935 a number of attempts were made to reconstruct damaged organs using implants of human or animal bone and cartilage. In 1936 the first success was achieved using rib cartilage. This is not as unusual as it might first seem since many mammals have a bone in the penis. In 1950 the first successful artificial implant was used and in 1960 such implants were improved with the use of various perforated acrylic materials by Dr. Pearman (Gee 1975, pp. 404–5). "Pearman's Penis" inaugurated the charming tradition of naming artificial penises after their inventors. These implants were not always successful,[5] and even when they were, some recipients complained about always being in an erect state.

The next great advance in penile technology was a vacuum penis that could be inflated by squeezing a small pump hidden in the scrotum and deflated by pressing a release button on the pump. "Scott's inflatable prosthesis" as Dr. Domeena Renshaw named it in an editorial in the *Journal of the American Medical Society,* "confirms that phallic worship is alive and well in the United States today" (Renshaw 1979, p. 2637). Dr. F. Brantley Scott and his team at Baylor Hospital in Texas designed their "implantable, inflatable prosthesis" as a modification of a "implantable, dynamic, artificial sphincter" and had it manufactured by American Medical Systems. They promised Scott's Prosthesis would have a 20-year life span based on their laboratory tests (Scott et al. 1979). Some of the psychological effects of these dynamic organs will be described below in the section on the psychology of cyborgs.

THE ARTIFICIAL KIDNEY—FROM PROLONGING DEATH TO PROLONGING LIFE

There's a difference between prolonging death and prolonging life.
—Dr. DeBakey (Fox and Swazey 1974, p. 174)

The technical result of the treatment was satisfactory, but the patient died the following day.
—Dr. Alwall (1986, p. 94)

The artificial kidney is the first successful artificial organ. Originally, however, it seemed only capable of prolonging death. While research on artificial kidneys can be traced as far back as 1913 to work at Johns Hopkins on a hemodialyzer (blood filtration) system, a successful artificial kidney was not made until 1944 when W. Kolff made his first rotating-drum dialysizer in the Nazi-occupied Netherlands (Kolff 1979a; Alwall 1986, pp. 86–8). Almost all of Kolff's first patients died but eventually his work, and that of others, especially Dr. John P. Merrill, led to the development of effective dialysis (Kolff 1979a). In the early days of dialysis other treatments of uremic poisoning offered a much better success rate than kidney machines. The *British Medical Journal* of October 1949, for example, compared the 50 percent success rate of a diet treatment with the 5 percent success rate of hemodialysis. That year the *Yearbook of Medicine* challenged the wisdom of working on artificial kidneys and in Great Britain and the Netherlands hemodialysis was suspended (Alwall 1986, p. 93). In this case, proponents of artificial organs were correct in their prediction that their work could lead to the most effective treatment of kidney failure. Interestingly enough, as Nils Alwall, a famous Swedish developer of artificial kidneys, admits, it was surgeons, attracted to the chance to get involved in kidney treatment, who strongly supported the development of dialysis over "strong opposition, especially from colleagues of internal medicine" (Alwall 1986, p. 93).

While some effort has gone into making portable artificial kidneys, most dialysis involves having the patients attach themselves to the kidney machine for some period of time. This kind of intermittent human-machine symboisis could be termed a *semi-cyborg*. This category of cyborg is bound to increase as home dialysis becomes more popular and as other techniques for feeding, medication, or blood cleaning (such as with artificial livers) continue to proliferate. Unlike a pure cyborg where the artificial body part(s) are permanently attached, semi-cyborgs are perfect examples of the spreading cyborg society which involves the simultaneous and progressive linking and unlinking of humans with various machines.

This traffic between the organic and the inorganic also includes animals, on both the embodied cyborg and the cyborg society levels.

THE ARTIFICIAL LIVER—MAN-ANIMAL-MACHINE SYMBIOSIS

The artificial liver research . . . started as a dreamful challenge.
—Motokazu Hori (1986, p. 211)

We are not talking about one species here . . . [O]nce you cross that [human-animal] barrier and are able to use animal organs, then the technology inevitably—as it always has anyway—is going to move forward and more and more ways of cracking [the species barrier will be found].
—Dr. Thomas E. Starzl (Altman 1992b)

Using animal organs or tissue in humans is a growing part of the contemporary proliferation of cyborgs.[6] It is a question of boundaries and borders and cyborgism can in many ways be seen as a full scale assault on traditional divisions (such as machine, human, animal) with an inevitable Balkanization and proliferation of organisms including many different types of cyborgs and other monsters (Haraway 1990a). When animal organs or tissue (or even tissues and organs from other humans) are used for transplants, that tissue is conceptualized as a machinic element, not as organic matter, and it is integrated into the machine-body of the recipient using the whole array of tools and attachments. Certainly software (the immune system) has to be modified or disabled for these repairs to take place,[7] but in terms of discourse *healing* is not the descriptive metaphor; *repairing* is. Unsurprisingly, as the "programming" of organic immune systems becomes easier, the number of xenotransplants will certainly increase. They could easily surpass the number of human transplants now performed (over 10,000 in the United States in 1989) because the major constraint on human transplantation is the availability of organs.[8]

None of the 33 known attempts since 1905 at using animal organs in humans has yet been successful. The most famous implant was the "Baby Faye" case in 1984 where a baboon heart kept a small girl alive for 20 days.[9] In 1992 a baboon liver was given to a man at the University of Pittsburg medical center by a team led by Dr. Thomas E. Starzl, in the first of what is expected to be four trials of a new anti-immunization drug known as FK-506 combined with three older drugs (Altman 1992a, A1).[10] If the new regime of immunosuppressive therapy succeeds, the door will be open for the widespread transplantation of various animal organs, especially from pigs and baboons, into humans (Altman 1992b, c).

Motokazu Hori, a leading researcher on artificial livers in Japan, notes in his internalist history of the artificial liver that there have always been three "streams of development" of the artificial liver: the biological, the artificial (here he means inorganic), and the hybrid (Hori 1986, p. 212). Because of the complexity of the liver, the biological and the hybrid approach have been the most successful. Nonorganic blood purification techniques sometimes return patients to consciousness from their hepatic comas but they do not save the patient's life because they cannot perform various metabolic functions only living liver cells can manage as of yet (Hori 1986, p. 212). Hori concludes somewhat apologetically that:

> Because it is not possible to play the 'Almighty' in the development of a complete artificial organ whenever a difficulty is encountered, a way must be found to apply the concepts and specific technologies of a hybrid system. (Hori 1986, p. 212)

Indeed the first artificial liver (1956) was a hybrid arrangement where a human was hooked up to a living dog's liver "incorporated in a cross-hemodialyzer" (Hori 1986, p. 211). And every successful artificial liver system since has also been of a hybrid form, almost all of them developed in Japan, which in 1979 launched a national project to develop the artificial liver.

The direct use of animals for organ transplants or in hybrid artificial organs is only a deepening of their role as research animals. Animal experimentation has been, and still is, crucial to artificial organ research. Rabbits, sheep, pigs, calves, dogs, rats, hamsters, and others have all been used to test and retest various artificial organs. Calves have been especially central to artificial heart research and many of the prototype hearts were, and are, judged on how long they can keep the calf alive. Such work, by necessity,[11] often seems cruel, such as the literal explosion of calves whose artificial hearts produce too much pressure but to make the hearts, such tests are crucial. (Foldy 1992, personal communication).

Animal research and xenotransplants demonstrate that cyborg medical solutions are not free in terms of organic suffering. They are not free economically, psychologically, or politically either.

THE ARTIFICIAL HEART—THE COST OF CYBORG SOLUTIONS

[T]he mechanical heart borders on science fiction and wishful thinking . . ."
—Dr. Denton Cooley, nine months before performing the first human artificial heart implantation into Haskell Karp. (Fox and Swazey 1974, p. 153)

On or about April 8, 1969, Haskell Karp died as a result of surgical experimentation performed on him by Dr. Denton A. Cooley, Dr. Domingo S. Liotta, and others in St. Luke's Episcopal Hospital in Houston Texas . . . Defendants, using devices approved only for animal experimentation, after removing the human heart of Decedent from his body, implanted an experimental mechanical device in Haskell Karp . . . The mechanical heart, in fact, had not been tested adequately even on animals, had never been tested on a human being . . .
—Mrs. Haskell Karp (Fox and Swazey, 1974, p. 205)

Because of its mythological and metaphorical resonances, the artificial heart has been given the most publicity, even though its medical impact has been significantly less than that of the artificial kidney and other technologies.

The first use of a totally implantable artificial heart (TIAH) was in 1957 at the Cleveland Clinic. Willem Kolff and Tetsuze Akutsu put two compact blood pumps in place of a dog's heart. The dog survived for about ninety minutes (Dutton 1988, p. 92). Since then there has been continual research on TIAHs including literally thousands of operations on calves, pigs, sheep, dogs, and people using dozens of different designs. Much of this work has overlapped with the development of different "assist" devices, especially those for the heart's left ventricle which does most of the pumping work. Various power sources have been considered including nuclear radioisotopes to produce steam[12] and the body's own skeletal muscles. Lithium and other improved batteries seem to have won this particular contest. There has also been a great improvement in blood compatible artificial heart materials and in artificial heart valves, which have benefited many thousands of heart patients even if they haven't yet led to a perfect TIAH. Most recently miniaturized control systems have been developed to synchronize the heart to the body's needs, either through biofeedback, outside programming, or both.

Much of the funding for artificial heart research has come from the federal government, especially the National Institutes of Health (NIH).[13] Independent medical nonprofits, foundations, and university medical programs have also devoted significant resources to artificial heart research as have several medical conglomerates and startup companies often formed by the artificial heart researchers themselves.[14]

Decision making around the artificial heart has been marked by a great deal of economic calculation as well as political and psychological factors. As Michael Strauss details in his article, "The Political History of the Artificial Heart," the artificial heart itself came out of the biomedical revolution which "was in large measure political in nature" and driven by "an alliance of interested congressmen, powerful NIH directors, and eloquent science lobbyists" (1984, p. 333). By the late 1970s the project to produce an artificial heart "had acquired a degree of autonomy and a momentum that was both scientific and political" (p. 336). It is a dynamic that is still in force today.

Economic Costs

There are two economic issues of note. First, the money spent on the artificial heart could save many more lives if spent elsewhere, especially on preventive medicine or increasing medical services among the impoverished (Lubeck and Bunker 1982; Dutton 1988). While artificial heart advocates have often tried to defend the program with straightforward (and unusually crude) calculations about the supposed benefits of artificial hearts in terms of returning x number of people to the economy (Kolff 1977; Kantrowitz 1983), they avoid, for obvious reasons, comparative calculations about the effect of the resources spent. If resources were infinite, a crash program to develop the artificial heart might make sense.

The second economic factor is the possibility that artificial organs will represent a very profitable product, hence, the proliferation off startup companies (see note 14). Even if there are no immediate profits, the benefits of publicity make the promotion of spectacular artificial organ research well worth the effort. For example, Humana, Inc. went out of its way to recruit Dr. William DeVries, the University of Utah heart surgeon who performed the famous operation on Barney Clark, for its own research center. After he performed his first transplant for them, the number of enrollees in the chain's health maintenance organization more than doubled in the next six months, leading *Business Week* to point out the hiring had "already paid off in spades" (Dutton 1988, p. 124).

Humana's president, William Cherry, insisted that profit making was not their incentive. "My heart is as ethical as anyone's heart," he promised. But looking at DeVries's move from the University of Utah to Humana it seems obvious that ethics were not the main motivating factor. DeVries had become disenchanted at Utah because the Institutional Research Board (IRB), which must approve all experiments with human subjects, was somewhat critical of the Barney Clark transplant on several grounds.[15] Even though the Federal Drug Administration (FDA) had approved six transplants at Utah, Utah's IRB was only approving them one at a time. The chairman of Humana, David Jones, asked DeVries during his recruitment, "How many hearts do you need to find out if it works? Would ten be enough?" When DeVries said that it would, Jones replied, "If ten's enough, we'll give you 100" (Dutton 1988, p. 123). Soon after moving over to Humana, Humana's IRB gave DeVries permission to perform six more implants.[16]

Psychological Costs and Political Pressures

The major motivator for artificial heart research is probably death. As Willem Kolff said,

> People do not want to die, and no government committee, no FDA regulations, no moral or theologic, not even economic considerations are going to change this. (Kolff 1979b, p. 12)

Death motivates funders, especially geriatric politicians with heart disease (Dutton 1988, p. 95); it motivates physicians who don't want to give up on patients and probably fear death as much as anyone[17]; and it certainly motivates many of the patients who turn out to be the experimental subjects.[18]

The desire for "fame and fortune" is probably just as important among artificial heart researchers as is any desire to conquer death, as can be seen by DeVries's move to Humana. Dr. Denton Cooley, now of the Texas Heart Institute, offers an even clearer example. His lust to be the first surgeon to implant an artificial heart in a human led him to violate numerous NIH and other regulations and

Medical Cyborgs 153

even scheme to appropriate a heart from another research team in 1969. All so he could thrust it into Mr. Haskell Karp's chest, ostensibly as a "bridge" until a transplant heart could be found but actually so Cooley could become a footnote in history. There is very little doubt that Cooley did it for fame. Mr. Karp died after three unpleasant days with an artificial heart and one with a transplant heart. The donated heart was wasted on Mr. Karp in any event as the artificial heart had practically killed him.[19] As Diane Dutton notes, Cooley's actions were, "to understate the case, ethically dubious, as there was no reasonable expectation of scientific advancement or benefit for the patient" (p. 103).[20] Cooley was forced out of his position at Baylor University by Michael DeBakey, a more principled researcher and the leader of the team that produced the heart Cooley hijacked. But Cooley's career did not suffer.[21] He moved full time to his current position and he even received a standing ovation from his peers at the next annual meeting of the American Society for Artificial Internal Organs (Fox and Swazey 1974, p. 176). Over ten years later, after learning that the University of Utah team was going to do an artificial heart transplant soon, Cooley rushed into another one, ostensibly as a bridge for a transplant. His patient suffered a week (two days with the artificial heart) and then died. Again Cooley failed to gain approval of the local IRB and he also violated new regulations of the FDA for using a new medical device without their approval. He was given a warning by the FDA which he mocked in public (Dutton 1988, pp. 114–5).

It should be stressed that no other artificial organ researcher has so nakedly displayed his disregard for his patients' welfare, although others have violated FDA and other regulations (Dutton 1988, p. 115). But Cooley's hubris and drive are present in lesser amounts in all of the researchers, as most will admit. One informant told some researchers on the Cooley controversy,

> [A]nybody who gets involved in the artificial heart has to be a very ambitious man ... because if things work out well [it holds forth] the possibility of instant immortality. (Fox and Swazey 1974, p. 188)

Using artificial hearts as a bridge has had some success recently,[22] and the work to perfect it continues. Many people live as full medical cyborgs, either linked to immovable machines temporarily or permanently, or with implanted cybernetic systems. Looking at their experiences offers interesting hints about our future as cyborgs.

THE PSYCHOLOGY OF CYBORGS

According to Bruce Hilton, director of the National Center for Bioethics, an incredible "70 percent of hospital deaths now involve the decision to withdraw some kind of support" machinery (he objects to calling it "life" support) (Hilton 1992). Clearly, for the terminally ill, let alone those without brain functions, cyborg medical machinery is problematical. Many people who have lived symbiotically with kidney machines, for example, reject the option of cardiopulmonary resuscitation in case of heart failure (36 percent in Tzamaloukas et al. 1990). The major factors determining their decision included their economic condition along with their general age and health. Many poorer patients apparently don't want to prolong their lives with what doctors often call "heroic" measures. Heroic for whom? one has to wonder. Many other terminal kidney patients who exist through dialysis machines terminate their treatment on their own (8 percent suicides reported in Van Neiuwkerk et al. 1990).

Willem Kolff (1989), aware of this problem, has advocated deferred consent among artificial heart recipients so that patients who wake up and are disturbed when they hear the "clicking of the pump" where once was the beating of their heart, could not disconnect their mechanical pumps until twenty-four hours had passed (p. 184). To put the problem of suicide in perspective, he reminds his colleagues that "there are also people with normal kidneys and normal hearts who commit suicide because they cannot cope with life" (p. 183).

Numerous studies (Devins et al. 1990; Petrie 1989; Zimmerman 1989; Kjellstrand et al. 1989; Gonsalves-Ebrahim and Kotz 1987) have shown that kidney transplant recipients do better than patients on dialysis and that those patients who can successfully use home dialysis do much better than those who must use the hospital. Clearly, the more normal the patients' lives can become the happier they are. Even life-sustaining machinery is sometimes rejected if the quality of the patient's life becomes too low, either for physical or psychological reasons, or for both. As Zimmerman (1989) notes: "Therapeutic dependence on technical devices represents a major intrusion into the life of the patient." Other factors that seem important are 1) the age of the dialysis patients, the older the patient the less they seem willing to bear to continue their lives; 2) whether or not they continue work; and, significantly 3) their gender. Several studies (Wolcottt et al. 1988; Gonsalves and Kotz 1987) indicate that men have more difficulty than women in adapting to an invalid sta-

Medical Cyborgs

tus, especially in terms of role reversals (from provider to dependent) and the body image distortion involved.

Many medical cyborgs go through several stages coming to terms with their mechanical prosthesis and that mechanical systems can be liberating as was shown by Joseph Kaufert and David Locker (1990) who followed the rehabilitative experiences of ten victims of the Manitoba poliomyelitis epidemics of the 1950s.

The patients followed a number of different "careers". First came the "acute" phase, the fight to survive. It involved tremendous physical effort and reliance on the iron lung for survival. Then came "rehabilitation" where dependence on the iron lung and other machines was seen as an impediment to independence. Then came "stability" which also meant constant exercise and conscious breathing to minimize or eliminate reliance on medical machinery. As many of these patients grew older and entered a "transitional" stage marked by medical setbacks, newer equipment, especially portable respiration machines which could fit on wheelchairs, were developed. Because of this technological change, patients who could not maintain themselves without machinery could still increase their independence with the help of portable machines; although before, reverting to the machine meant losing independence. Even for patients who could breath on their own (albeit with great effort), the portable respirators offered an increased quality of life.

Ironically, this independence came at the expense of increased reliance on machinery (Kaufert and Locker term it "partial dependence") and on other friends and family who have to be prepared for mechanical, and therefore medical, emergencies which would fall on the hospital staff if the patient remained there. The extreme work ethic of the rehabilitation and stability phase also had to be overcome, since it was linked with a rejection of machine dependence. As Kaufert and Locker note: "Within rehabilitation ideology virtue lay in resisting the temptation of the machine: dependence on equipment rather than the self was depicted as idleness and having given in to the disease" (p. 872).

Kaufert and Locker stress how important the respiratory machinery is to the patients in the machine-dependent stage in terms of new costs and new benefits.

> Some of these costs and benefits were attributable to the new body/machine interface, while others were indirect and the result of a changed relationship to the physical and social environment and to the self. In this regard the machine had both practical and symbolic impacts . . . (p. 874)

The most difficult surrender, for some patients, was accepting the creation of a permanent tracheal airway for the portable respirator. It necessitates quite a change in "body image" because it involves a "direct physical connection to a machine." However, the patients who went through with it often considered it a "small price to pay." As one woman put it:

> I think that anybody that feels that they don't want to go on the mini-lung because of having to have the trach and having to have the house (sic.) connected, they don't realize what freedom they've got. (ibid)

Accepting the machine is not a simple process. The patient has "to adapt to the rhythm of the machine." As one stressed, you have "to learn the machine. Nobody can tell you, nobody can teach you, you have to learn the machine." Even more, the patient has to live with great awareness of the machine for the rest of their lives or until they return to the hospital. They have to listen to its functioning and feel its operations (the air pressure and vibrations) and be prepared to repair it or adjust it if it malfunctions.

> As a consequence, the machine, and tending to the needs of the machine, become the central focus and activity of everyday life. Moreover, the problematic aspects of living with the machine did not detract from its role of promoting a quality of life of more than minimal tolerability. (Ibid)

The machine becomes part of even the most intimate moments of the patient's life. One recently married woman describes what this specifically meant to her and her husband.

> When we received our marriage vows the doctors didn't tell us, they didn't know themselves, about the intercourse, that your respirations increase. The machine just gives you your ten or eleven breaths per minute, so I need sixteen breaths per minute during that time I need more air. My husband and I had to interpret this as another experience—turn the rate button up, increase the pressure, turn this up, and he was paying more attention to the respirator than he was with . . . and trying to keep his mind on everything. All of a sudden as you're enjoying intercourse you have to turn the rate button up. He's had to learn to do all these things. (Ibid)

Still, humans are adaptable. Kaufert and Locker stress that,

> All those respondents who had made the final transition to a ventilator with a tracheal airway readily agreed that it had a positive impact on

Medical Cyborgs

their lives. The differences in quality of life were so dramatic that portable ventilator users often tried to persuade other post-polios to make the transition too . . . A transformed sense of self accompanied the new dependence upon machinery and the changed relationship to the environment and the people in it. (p. 875)

Choosing a cyborg existence for these people clearly has made them more independent. Overcoming traditional fears of machinery (dependence on and linking to) and their learned rehabilitation ethic allowed them to become free of the hospital and free to become active citizens again. As Kaufert and Locker concluded, this new ideology about cyborg machinery was as important as the medical condition of the patients themselves.

The nature and extent of the gains and losses experienced by the respondents appeared to be influenced by variation in individual functional status and degree to which people were able *to successfully modify culturally grounded beliefs about dependence upon technology.* (p. 876, my italics)

It is significant that one of these patients became a leading advocate for disabled rights in Canada after she adopted the portable respirator. This transition, to independence through dependence on improving cyborg technologies, is not unique and is clearly a major factor in the growing disabled rights movement (Crewe and Zola 1983).

The polio victims were confronted with a new technology after years of working to free themselves from an older, more intrusive and limiting technology. In the case of amputees, there is the established culture of artificial limbs available from the beginning. It seems with them, there is a good chance of retaining their bodily image with little change.

In one of the first major studies of how wearers of prosthesis felt about their artificial limbs, a surprising number (61 percent) claimed that they forgot they were an amputee "most of the time" and 7 percent even claimed they forgot "all of the time." Careful psychological testing revealed that these numbers were somewhat exaggerated but that humans did have a tremendous ability to adapt to their new situations by maintaining feelings of "bodily integrity and adequacy" through denial (Silber and Silverman 1958, pp. 91, 115). More surprisingly, almost half the amputees claimed to be able to do "as much" as nonamputees and 14 percent claimed to be able to do "somewhat more" (p. 92). Feelings of superiority and inferiority by the amputees over nonamputees were almost the same (p. 99).

In what the authors considered a similar pattern of denial, 61 percent of the amputees said that while hooks were "mechanical-looking" they were not unsightly, while 13 percent said they were "as natural looking as any hand" (pp. 95–6).

> The amputee's preferences in artificial limbs, and his habits in using them, are evidently not based entirely upon his objective assessment of his functional and social needs. They are influenced also by emotional factors arising from the meaning he attaches to the wearing of artificial limbs. (p. 99)

This demonstrates once again that body image and cultural attitudes toward technologies are relatively plastic and certainly constructed and reconstructed in society and subcultures (in these cases medical and specifically patient communities) again and again.

Another major group of medical cyborgs that has been studied are men who have had mechanical inserts installed in their penis to cure impotence. A large percentage of these men don't reveal to others that they received an implant, even to their sexual partners (Tiefer et al. 1988, p. 189). One patient admitted that when his sexual partners asked him why his penis was so large he'd always tell them "It's just the way you make me feel" (Tiefer et al. 1991, p. 118). In one study (Tiefer et al. 1988), almost 20 percent of recipients even tried to hide their penis from strangers while using restrooms, since with a rigid implant it is semi-erect. Somewhat more surprising, 10 percent become exhibitionistic, some even partially inflated their vacuum implants before going out (p. 190). While a fair amount of satisfaction was reported in many studies (50 percent in Gee et al. 1974; 80–90 percent in Coleman et al. 1985, p. 199), more careful research that interviewed partners as well as patients revealed that actual rates may be much lower (Tiefer et al. 1988). For example 40 percent of one study reported some malfunctions (Tiefer et al. 19091, p. 113). Dreams about the prosthesis "blowing up" (p. 121) and anxiety about having a foreign object in the body or having it break (p. 118) were common. Even one patient who was very outspoken in public about his vacuum prosthesis displayed, according to Tiefer and her colleagues, a great deal of distress. "He seemed to relieve his anxieties through bragging, telling other men to 'eat their hearts out' and saying he was called the 'pump man'" (p. 124).

It is also significant that, "Few patients were able to describe precisely how the prosthesis worked or what actually was implanted or changed during the surgical procedures" (p. 117). Ignorance among the patients in this study group, who were all going through corrective surgery because of penile prosthesis malfunctions, was extraordinary and it is in marked contrast to the attitude of the polio patients who use artificial respirators or of home dialysis patients.

For contemporary medical cyborgs the technologies in question are viewed somewhat pragmatically, and ideologies are constructed accordingly. Quality of life is subjective but, as the different decisions of all of these patients show, it is one concept that is acted upon. The plasticity of attitudes toward human-machine integration is of particular note in this regard. Machine integration and a declining quality of life might lead to suicide while machine integration and an improving quality of life clearly can lead to growing independence and activity. However, cultural attitudes do not disappear, as the responses of penile prosthesis users reveal.

With medical cyborgism the patients are probably the main driving force behind the development of new cyborg technologies, but nothing would be done without the doctors, the bureaucrats, and the businessmen. And I use "men" advisedly. Of the hundreds of key players I have seen named and described in the research, *none* are women. Women play a role in judging the social impact of the technology and even in psychological studies, and certainly they are present as patients and wives as well as nurses and technicians. But as doctor-inventors, surgeons, NIH bureaucrats, or corporate leaders they are almost entirely absent. This gendering of the actual invention and the surgical interventions of the body is one that comes as no surprise to anyone familiar with feminist explorations of the patriarchal underpinnings of contemporary western science and technology (see Haraway 1990a for an introduction).

THE DESIRE FOR CYBORGS—ECONOMIC AND PSYCHOLOGICAL

> Commitment to the design, construction, and implanting of artificial internal organs requires a positive, romantic, and unrestrained view of what may be attainable. Members of our Society [International Society for Artificial Organs] share a bond gained by the belief that fantasy can be transformed to reality.
> —Dr. Eli A. Friedman, president ISAO (1987a)

There are many causes for the current proliferation of medical cyborgs. It is over determined, as so many technocultural processes seem to be these days. These causes are both personal and social, and it is important to stress that personal decisions have a real effect. Modern medicine is usually written as a history of "great" (male) doctors but there is some truth to this perspective, especially as western medicine is a subculture of incredibly masculine assumptions and rules. By masculine I don't mean just male. Aggressiveness, competition, risk taking, physical and technical skill, cutting open living bodies, making life and death decisions, making lots of money, having a great deal of social power and prestige, are all qualities that are gendered masculine in our culture but they are not biologically linked to men. Women can, and do, perform these tasks very well as men can also take orders, get paid less, and be caring, nuturing, and emotive, which are qualities gendered feminine and are relegated to nurses in medicine. By gendering surgery and invention masculine, two things are accomplished: 1) power is kept in the hands of men by and large; and 2) qualities gendered feminine are kept from being important in the discourse. This gendering plays a real role in the dynamics of cyborgism as can be seem by a listing of the basic forces behind it.

Justifications for the development of medical cyborg technologies can be grouped under the following basic categories: social and personal.

Social Justifications

Regarding economic reasons, the most salient are the increased incomes of doctors, the profits of medical companies, and the general economic health of the country. Note that even though the national economy would be better served by more preventive (more feminine) medicine, high-tech and surgical interventions are always defended on economic grounds.

In masculine discourse social standing is crucial. So it should come as no surprise that cyborg technologies are often defended in terms of competition, whether or not it is between countries (Kolff 1989), between institutions, or between doctors, as with the competition for "firsts" around the artificial heart. The military's interest in cyborgs is certainly institutional but it is driven by instrumentalist efficiency values. Such values don't seem as important to medical institutions except as they relate to the corporations that make artificial organs and other supplies. There the similarity between corporations and the military is striking since medical suppliers of the latest cyborg technologies often compete on quality since, in many ways, cost is not a part of market determination (unless products are judged absolutely equal). Who would advocate buying an inferior artificial heart (or robot weapon) just because it was cheaper?

Personal Justifications

Healing is often mentioned as the major motivation by doctors and it is certainly a primary interest of patients and their loved ones. The doctor's drive to heal is not framed as nurturing, however, but as solving a problem. Still, in most of the cyborg inventors, it seems a very real motive, and references to their feelings of helplessness as they watched patients die are common. Equally sincere is their anger at attempts (necessary attempts actually, see Smith 1990) to make cost benefit determinations about who will get what kinds of medical care. For a doctor with a specific patient, such a determination is frustrating in the extreme, as Willem Kolff makes clear in his inimitable style:

> The Alte Pinakotek in Munich has a large and extraordinary collection of paintings. The enormous painting by Rubens, "The Last Judgment," shows the Lord and Christ sitting on clouds looking down on the crowd of damned naked women that try to struggle upwards from Hell, while being bitten by all kinds of monsters. It reminded me of committees instituted in some places to decide which patients in chronic renal failure may be treated with the artificial kidney and which may not. (Kolff 1966, p. 328)

It is very significant that many of the cyborg inventor-doctors (such as Kolff and John P. Merrill who served as flight surgeon on the *Enola Gay*) have been very active in physician groups organizing against nuclear war. Their commitment to life cannot be questioned, even if it may be a bit out of balance. However, they have other desires as well.

Hubris and the desire for fame are motivating forces. This applies both to doctors and medical administrators.[23] Consider what Dr. Eli A. Friedman said of his great friend John P. Merrill, "America's First Nephrologist":

> Merrill was cognizant of his place in history. Like Hemingway, who was his friend. JPM—as he was known to his students—constantly pushed the limit of his physical and intellectual endurance. Whether searching for the bigger marlin off Cuba, the more menacing lion in Kenya, or the perfect dose of total body radiation in Boston, JPM was electric to observe ... Merrill had class. Selecting the correct wine for a visiting Frenchman was regarded as a challenge ranking only slightly below the choice of dialysis time for a newly referred uremic patient. (Friedman 1987b)

This passage abounds in masculine themes, especially of hunting. And as for comparing choosing wine to choosing therapy, it certainly reveals a high level of objectification and competition.

The pride and confidence of these men can be seen in the epigraph from this section that admits how much the creators of artificial organs are driven by the desire to make fantasies real. Dr. Eli A. Friedman, president of the International Society for Artificial Organs (ISAO) deserves to be quoted at more length. In the following passage he describes the connection between science fiction and science in an editorial in the ISAO's journal *Artificial Organs,* entitled revealingly "ISAO Proffers a Marvelous Cover for Acting Out Fantasies."

> While no data substantiate the point, I consider it safe to speculate that a high proportion of ISAO members were enthralled by science fiction as restless adolescents, and still recall how Wells, Heinlein, Van Voght, and Asimov told of a world in which imagination was unrestricted. Future ISAO members walked on other planets, regressed or advanced in time, and watched as the dead were restored to health by cloning individuals from cell parts. A recurrent theme in 'SciFi' is the acceptance of medical facilities termed 'cosmetics' in which severely wounded Starpersons receive spare parts permitting normal vigor. As we near the close of this century, many of these hopes for tomorrow have been absorbed into daily life. (Friedman 1987a)

Dr. Pierre Galletti (1986), echoes this emphasis on the transendence of the now with the invention of the future when he reprises George Shreiner's categorization (from his presidential address to the American Society for Artificial Internal Organs in 1959) of different types of artificial organ developers into three categories. They are "inventor," the "harvester," and the "gleaner." Now which of these has the highest status? Galletti does make the important point that patients have become very important in artificial organ research because of their observations, and because many of them have become researchers themselves (p. 386).

The cyborg inventors and surgeons often talk of "winning" (see Altman 1992c for example) but it isn't always clear who they wish to win against. Is it death? Other doctors? People who doubt their technologies or their skill? Themselves? My impression is they mean all of them. Life is a competition, death the ultimate victor, but the game will go on.

Self-interest is certainly an important personal factor. There is the self-interest of career, and there is always the very human and widespread fear of disability and death. Many artificial organ re-

Medical Cyborgs

searchers started as patients (see above); this is true of artificial limb developers as well.

> Every prosthesist knows of amputees who are always looking for something better. Sometimes such persons channel their needs constructively and make a contribution by entering the field of prosthetics development. More often, however, they dissipate their energies going from limbshop to limbshop looking for satisfaction they probably cannot get. (Silber and Silverman 1958, p. 110)

The fear of disability and death can be overcome by rationalization and conscious effort. In an article on careers in bioengineering, one writer notes that by thinking the right thoughts fear can be conquered.

> Confronted in an experimental surgery room with the scene of a doctor holding a dog's pulsating heart in his hand, an engineer might well be intimidated (at least the first time he witnesses this event). But given some kind of system diagram with which to think things out, and to play around with, his interest (in the "problem") and his hopes (that he might solve "it") may revive and overcome his apprehensions and doubts. (Lindgren 1965, p. 73)

The medical cyborg inventors are so focused on their goal that they don't consider the complex ethical and social implications of their work. Willem Kolff has even gone so far as to claim that ethical studies of artificial hearts are just an "excuse to delay the spending of monies" and he promises that "no light will emanate" from such reflection" (1989, p. 184). Another manifestation of this is found in the continual hyperoptimistic predictions (so similar to computer scientists). For example Dr. Nosé predicted in a 1979 interview that TIH would be possible by 1986 (Foldy 1979, p. 1). There are numerous other examples.

A key part of the psychology of medical cyborg-makers is the confusion between the human and the machine. In part it can be traced to the victory of biology (biomedicine) as the dominant metaphor for contemporary western medicine after several hundred years of struggle with more religious conceptions. Dr. Seth Foldy argued in 1978 that this meant that there was a shift in the "basic underlying principle of medicine—from quality to quantity, from form to machine," He went on to note that,

> Whereas traditional Galenic medicine had embodied at its deepest level concepts of goal and purpose, of essences and values, the mechanistic

model that followed it was rendered relatively devoid of these qualities. Medical science was cut off by the "arbitrary separation of values" from an outlook that included *positive* (if metaphysical) *potentials* that had been perceived beyond the *existent, empirically measurable* quantitative facts. In the process, the scientific understanding of the human being had shifted from a value-laden form to the object of empirical investigation, epitomized in its interchangeable, measurable, manipulable qualities in the form of a dissected cadaver. Organs could be removed, studied and replaced. (Foldy 1978, pp. 2–3, original emphasis)

The biomedicine discourse is now being challenged by a new computer-driven metaphor of "infomedicine" which is based on the "postmodern sciences" of quantum mechanics, irreversible thermodynamics, and, of course, information theory (Foss and Rothenberg 1987).

The influence of this infomedicine model is clear in this statement by the eminent British doctor and prosthesis engineer D. S. McKenzie.

> One of the greatest virtues of biological systems is that they are highly adaptive. The human control system—and in particular the computer as represented by the central nervous system—is no exception to this. (McKenzie 1967, p. 3)

Jean-Louis Funch-Bretano shows a similar commitment to the new epistemology of infomedicine when he argues that "the artificial organ field is one of the medical fields where application of computer science could be the most fruitful . . ." He goes on to define expert systems as "genuine prosthesis for intellectual behavior for decision-making" (1986, p. 210). Despite his enthusiasm for informatics, Funck-Bretano does stress in his article that artificial organs are, as of yet, not as good as actual organs. Even as cyborg-makers try to improve artificial organs so they are equal to their natural equivalents, there is the related desire, among patients as well, to go beyond the norm: to improve on the human.

THE FUTURE FANTASTIC OF MEDICAL CYBORGS: THE IMMORTAL POST-HUMAN

> If scientific advancement continues to accelerate, there may come a time when every part of the body can be replaced, when no one will ever have to wear out, when humans can become immortal, at least in theory.
> —Gloria Skurzynski (1978, p. 132)

Medical Cyborgs

The possibility that artificial body parts could enhance humans, not just repair them, can be traced as far back as 1505 when Gotz von Berlichlingen, a German knight, lost his lower right arm and had it replaced by a prosthesis of metal that could be set for gripping a weapon or just used as a club. His effectiveness in his chosen occupation (pillaging) was by many accounts increased markedly by his "iron hand" (Skurzynski, 1978, p. 1). As he put it, his artificial hand "[h]as rendered more service in the fight than ever did the original of flesh" (Putti 1929, p. 111, no. 3, Putti's translation).[24] Even today, amputees find similar benefits in their limbs, as 10-year-old Josh Green, a Little League catcher from Elyria, Ohio, demonstrates when he uses "his artificial leg to stop wild pitches and block runners at home plate. 'It doesn't hurt me'" (Hudak 1992).

Many amputees have kept a set of artificial limbs, some for work and some show, to use as the occasion demanded. Gotz, for example, had more than three, including an ornamental wooden one and two metal mechanical hands of iron and copper for battle (Romm 1988, p. 159).

The Pump Man, mentioned above, was so convincing about the superiority of his vacuum penis that at least one coworker of his went to the doctors and asked for an implant even though he had no impotence problems (Tiefer et al. 1991, p. 126). Many penile implant recipients (50 percent in this study) felt the implants made the "feel more of a man" (p. 115). This even extended to the preference by some for the less realistic and more painful semirigid implants because they were "more masculine" (p. 116). Several of these men mentioned the attention they got from sexual partners, not knowing of the implants, who "marveled at their ability to maintain a constant erection."

The evidence of a widespread desire for greater-than-normal abilities from prosthetics is merely anecdotal, limited as it is to statements by patients and doctor-inventors, but one only has to look at human culture as a whole, with its fascination for gods, goddesses, monsters, regular and super heroes, and ubermench, to see how strong the desire for the posthuman might be. Add to this a desire for immortality with the evidence here that humans can certainly adapt themselves to machine-human symbiosis, and there is strong evidence for a powerful cultural bias toward developing cyborgs that go beyond the human, instead of just repairing them. There will certainly be lots of money to be made at it as well. Indeed, augmenting the human is the focus of almost all of the military and space research on cyborgs as well as a great deal of the more significant artificial intelligence research at some of the most elite institutions (MIT and Carnegie-Mellon, for example), where the dream of "downloading" human consciousness into machines is a serious research goal (Gray 1989, 1993). Cyborgism promises much to many different people.

THE PROMISES OF CYBORGS

Science has made us gods before we are even worthy of being men.
—J. Rostand (Cortesini 1984, p. 127)

The only deep unanswered question, after the technical problems are cleared away, is the question that will be posed by the man who is living only by the grace of an artificial heart, and whose "every second of life might be totally dependent on his local power company," namely, "**Who am I?**" Well may he ask. (Lindgren 1965, p. 83, original emphasis)

There are great promises in cyborgs, but are they promises to everyone? There is the very real possibility that the growing gap in health care between the well-off and the impoverished in the United States (Foldy 1990) will lead to two classes, rich cyborgs and poor humans. And it may also divide people along national boundaries.

The diffusion of dialysis technology is particularly revealing when one wonders about the racial, class, and national parameters of cyborgs today and in the near future. In Nils Alwall's history of the artificial kidney he notes that it wasn't until 1958 that dialysis was performed in Africa, and then it was only in Egypt. This was 15 years after the first successful dialysis in Europe and almost ten years after dialysis became widespread experimentally in Europe and North America. In his chart (Alwall 1986, p. 94, table 2) he lists sixteen sites of new dialysis experiments, all of which are European countries except for Australia, Egypt, India, and a strange nonnation category: "South America." His final chart shows that the vast majority of dialysis in 1981 occurred in European and North American countries, most others in Australia, Japan, and Israel.

Economic, national, racial, and cultural factors, such as attitudes toward machines and human-machine connections, could lead to other divisions between cyborgs and humans. Outright anger towards cyborgs is certainly possible. As the NIH's Working Group on Mechanical Circulatory Support warned:

Society's response to the recipient [of an artificial heart] is unpredictable. The recipient, his family and those in the health care environment may expect reactions ranging from unwarranted adulation to threats of bodily harm. (Van Citters et al. 1985, p. 391)

Fuller cyborgs *will be different* from humans.

> The recipient will probably have internal sensations—evoked by motion, impulse, and vibration—quite different from those evoked by a natural heart. External cues to the presence of the device will also be obvious through touch, sight, and sound. (Van Citters 1985, p. 391)

And an artificial heart, no matter how bad the pounding (and Clark hated the feeling in his chest—Berenson and Grosser 1984, p. 913), and how noticeable it is to outsiders (Clark's body shook with each beat—Gaffney and Fenton 1984, p. 918) is one of the least noticeable cyborg prosthesis. Animosity to the demands of wheelchair-bound medical cyborgs is already common, especially in industries most affected by their demands, such as transportation (G. Gray 1989, personal communication).

And while cyborgism may deliver longer life to some, and it may promise immortality eventually, it is already leading to a "living death" for others, such as dead pregnant women with living fetuses (Hartouni 1990). These cases reveal a serious danger of cyborgism: human bodies, already seen metaphorically as machines, can become nothing more than actual machines (in this case a hybrid artificial womb), especially if the humans happen to be from a group, such as women, whose status (legal and otherwise) is already partially defined by their reproductive function. Still, the mothers I know would gladly accept their already dead bodies being converted into a machine if it meant bringing a baby they wanted into life. Perhaps realizing that women (and men) can make such hard decisions allows Valerie Hartouni to end an article, where she has written eloquently about the danger of turning women into reproductive machines, with the bold statement that we should find a "certain courage" and

> ... take seriously the socially and technologically produced opportunity to invent ourselves consciously and deliberately, and in this to develop the practical, political implications of the philosophical claim that 'we' are only and always what we make. (p. 50)

We need such courage to fulfil the better promises of cyborgs and to resist the development of cyborgs merely to improve the killing efficiency of man-machine weapon systems or the profits of corporations. As Andrew Ross remarked while interviewing Donna Haraway, "It seems clear that there are good cyborgs and there are bad cyborgs, and that the cyborg itself is a contested location." (Haraway 1990b, p. 7)

Which brings us to a fitting conclusion, because it is Donna Haraway who has done the most to make others take the promises of cyborgs seriously. Certainly she is right that the dangers of cyborgs have to be weighed against what they offer. The threat of killer cyborgs stripped through drugs and surgeries of their human empathy and augmented electromechanically with weapons has to be balanced against the post-polio patient in Canada whose link to a machine has allowed her not only to live a better life, but to struggle to make Canada a better place, a more human place to use an honored metaphor, to live for everyone.

> A cyborg body is not innocent; it was not born in a garden; it does not seek unitary identity and so generate antagonistic dualisms without end (or until the world ends); it takes irony for granted ... The machine is not an *it* to be animated, worshiped and dominated. The machine is us, our processes, an aspect of our embodiment. We can be responsible for machines; *they* do not dominate or threaten us. We are responsible for boundaries; we are they. —Donna Haraway (1985, p. 101, original emphasis)

NOTES

1. I thank the National Endowment for the Humanities for their support at Case Western University for the summer seminar on "American Culture and Technology" during which most of this chapter was written. I also thank Prof. Carroll Pursell, our seminar host, for his graciousness and assistance. The librarians at Case Western and at the Allen Memorial Medical Library were kind and helpful as was Joanne Elser, director of the ISAO center in Cleveland. Finally, Seth Foldy, M.D., a dear friend, was invaluable in guiding me into the labyrinth of high technology medicine. If I've escaped with anything of value he deserves much of the credit.

2. A term coined by Manfred Clynes to describe intimate human-machine systems. See Kline and Clynes 1961, p. 356.

3. Which I owe to my friend and colleague Joe Dumit.

4. Reswick and Vodovnik, developers of artificial limbs, even offer a stricter definition: "When the orthotics or prosthetics system uses external power and is operated by means of feedback control the result is a cybernetic system in the true sense of the term" (1967,

pp. 5–6). However, external power is really not a necessity, especially as work to harness muscles and biochemical reactions to power artificial organs continues to make progress. Since cybernetics is an *information* science, what is important is the flow of information that feeds back and forth across the human-machine border, and energy transfers are not necessarily needed for that.

5. Gee et al. report (1974) that out of 19 rigid implants they performed in the early 70s only 15 could be considered a technical success and only 9 were sexual successes, as evaluated by the patients.

6. Other attempts at using animal tissue to ameliorate human conditions have been almost bizarre. Between 1889 and 1922 there were several attempts to use extracts from the testicles of animals, or even, as in the case of Prof. P. Lespinasse of Northwestern University actual slices of human or animal testicle, to treat impotence as well as more general maladies. These culminated in the work of L. L. Stanley, the resident physician at the federal prison at San Quentin, who injected 1,000 people, mainly inmates but some doctors and even seven women, with extracts of goat, ram, boar, and deer testicles (Gee, 1975, pp. 401–2).

7. Indeed, many of the earliest transplants were only successful because the recipients had been given massive irradiation treatments to disable their immune systems. A history of these interventions, and of the role of immunology in transplants is in Murray 1992.

8. In 1989 in the U.S. "there were 8890 kidney, 2160 liver, 1673 heart, 413 pancreas, and 67 heart-lung transplants" (Murray 1992, p. 1414).

9. Baboons are considered good candidates for organ donation as they are anatomically similar to humans, they are not an endangered species, and they can be easily bred in captivity (Altman 1992a, p. A12).

10. The patient lived less than three months. Dr. Starzl also performed the first human liver transplant. He also unsuccessfully transplanted baboon kidneys into six humans in 1963, one of whom lived 98 days, and later he put chimpanzee livers into three children, one of whom survived 14 days (Altman 1992c).

11. If there are to be artificial organs, then there must be such animal research. Not only must there be animal experimentation, but in a very real sense there must be human experimentation as is clear from the early history of the artificial kidney and of Barney Clark's experience with the artificial heart. It is easy to oppose these technologies in the abstract but not always in the flesh, as animal

rights protesters discovered when they protested the use of a baboon liver in a human patient only to be confronted with another terminally ill (unless he receives a successful animal liver transplant) patient (Altman 1992b). Is a baboon's life worth as much as a human's? Such a question shows that even with the proliferation of cyborgism there remains a hierarchy of value with the living human at the top and the cheap machine or technique at the bottom. Live animals, dead humans, dead animals, and expensive machines and techniques occupy shifting positions in between.

12. The nuclear heart was finally shot down by the NIH's Artificial Heart Assessment Panel which, among other concerns, noted that patients with plutonium in their chests would make good targets for criminals and terrorists and they might also sterilize or otherwise harm their bedmates (Van Citters et al. 1985).

13. The Atomic Energy Commission funded a great deal of research on nuclear power sources for hearts, that included at least one successful animal implant, until 1973 when political and technical problems convinced them that the nuclear heart was unacceptable (Dutton 1988, pp. 101, 104).

14. Such as Kolff Associates (now Symbion), which included among its major stockholders Willem Kolff, Robert Jarvik, Donald Owen, and the surgeon William DeVries, as well as the Hospital Corporation of America, American Hospital Supply Corporation, and Humana, Inc. (Dutton 1988, pp. 111, 115–6)

15. They were particularly concerned that a valve that failed in Clark's heart had not been tested, that Clark's lung disease had not been considered in choosing him, and that there were plans to try and get "healthier" patients so as to improve the chances of success with succeeding transplants. Their major issues were whether or not the health of the patient was being considered enough and whether or not there was proper informed consent. (Dutton 1988, p. 123).

16. He actually performed three within six months, none of which were successful (Dutton 1988, p. 125).

17. The desire to postpone death or defeat it entirely, seems to be a major motivation for many medical researchers and even for scientists in such seemingly unrelated fields as computer science, where schemes to eventually "download" consciousness into immortal robots play a surprisingly large role in the fantasy lives and research agendas of many eminent computer scientists (Gray 1993).

18. Although not entirely. Many patients apparently realize that they are unlikely to benefit directly from their experience. This is true of Barney Clark, a dentist. One of his sons said after Clark's

death, "He never really thought the artificial heart would work *for him*" (Dutton 1988, p. 118, original emphasis). The heroism of his sacrifice (and of his family's) is clear even through the haze of psychomedical jargon in Berenson and Grosser 1984.

19. To justify his actions Cooley staged a media call by Mrs. Karp for a heart donor and one was found in Lawrence, Massachusetts. Her heart was given to the dying Karp and her other organs were lost to other possible transplant recipients (Fox and Swazey 1974, p. 184–5).

20. None of the calf transplants had been successful. The one calf that had survived almost 13 hours was never able to stand and only had limited reflex movement (Fox and Swazey, 1974, p. 160).

21. Nor did Dr. Domingo Liotta's, Cooley's assistant, although he did lose his NIH funding he went to work directly for Cooley despite having also falsified animal data for at least one scientific paper (Fox and Swazey 1974, pp. 149–209). A law suit filed by Karp's family was unsuccessful, apparently because of extreme limitations on testimony permitted by the judge (ibid, pp. 204–10).

22. Johnson et al. 1991 report that between August of 1985 and October 1990 171 Symbion artificial hearts were used as a bridge to transplants and 69 percent of the patients lived long enough to get an organic heart, and 57 percent of those lived at least a year. One patient survived 603 days on the artificial heart. Between April of 1969 and October of 1990 217 artificial hearts of all types were implanted at 40 centers, with 69 percent either dying or being given an organic heart within two weeks.

23. And to experimental patients as well. Barney Clark mentioned that he wanted to leave a "small mark in this world." (Berenson and Grosser 1984, p. 911).

24. Romm (1988, p. 159) cites a latter source that disagrees with Gotz's self-assessment, however she also mentions at least one other soldier, a French army captain, who used an artificial limb effectively in battle (ibid, p. 161).

BIBLIOGRAPHY

ACM (Association of Computer Manufacturers) (1992) *Communications of the ACM,* issue on "Computers and People With Disabilities," Vol. 35, No. 5, May.

Altman, Larence K. (1992a) "Terminally Ill Man Gets Baboon's Liver in Untried Operation," *New York Times,* June 29, pp. A1, A12.

———, (1992b) "First Human to Get Baboon Liver Is Said to Be Alert and Doing Well," *New York Times,* June 20, p. B6.

———, (1992c) "Man Given Baboon's Liver in Transplant 'Doing Well,' Doctors Say," *New York Times,* July 1, p. A12.

Alwall, Nils (1986) "Historical Perspective on the Development of the Artificial Kidney," *Artificial Organs,* Vol. 10, No. 2, pp. 86–99.

Anbe, J., T. Akasaka, Y. Ogura, H. Nakajima, M. Ozeki, T. Mitsuishi, and K. Okinaga (1991) "Abstract: A Study on the Microcomputer-based Automatic Regulation of the Extracorporeal Circulation," *Artificial Organs,* Vol. 15, No. 4, p. 326.

Antypas, G., and J. Kavoukas (1987) "Abstract: Treatment of Tracheal Carcinoma by Tracheal Replacement," *Artificial Organs,* Vol. 11, No. 4, p. 342.

Berenson, Claudia K., and Bernard I. Grosser (1984) "Total Artificial Heart Implantation," *Archives of General Psychiatry,* vol. 41, Sept., pp. 910–16.

Broniatowski, Michael, Sharon Grundfest-Broniatowski, Charles R. Davies, Gordon B. Jacobs, Harvey M. Tucker, and Yukihiko Nosé (1991) "Abstract: Progress Towards a Dynamic Rehabilitation of the Paralyzed Face," *Artificial Organs,* Vol. 15, No. 4, p. 326.

Casper, Monica (1996) "Fetal Cyborgs and Technomoms," in Gray, ed. *The Cyborg Handbook,* New York: Routledge, pp. 183-202.

Chang, T. M. S. (1991) "Editorial: The 1991 ISAO-ISBS Congress and Artificial Cells," *Biomaterials, Artificial Cells, and Immobilization Biotechnology,* Vol. 19, No. 2, pp. v-viii.

Christenson, L., P. Aebischer, and P. M. Galletti (1987) "Abstract: A Bioartificial Thymus as a Potential Immunomodulator," *Artificial Organs,* Vol. 11, No. 4, p. 338.

Clarke, Adele (1996) "Modernity, Postmodernity, and Reproductive Processes," in Gray, ed. *The Cyborg Handbook,* New York: Routledge, pp. 139-56.

Coleman, Eli, Alan Listiak, Gordon Bratz, and Paul Lange (1985) "Effects of Penile Implant Surgery on Ejaculation and Orgasm," *Journal of Sex and Marriage Therapy,* Vol. 11, No. 3, Fall, pp. 199–205.

Cortesini, R. (1984) "What are the Ethical and Social Implications of Artificial Organs?" *Artificial Organs,* Vol. 9, No. 2, pp. 127–8.

Crewe, N., and I. Zola, eds. (1983) *Independent Living for Physically Disabled People,* San Francisco: Jossey-Bass.

Devins, G. M., H. Mandin, R. B. Hons, E. D. Burgess, J. Klassen, K. Taub, S. Schorr, P. K. Letourneau, and S. Buckle (1990) "Abstract: Illness intrusiveness and quality of life in end-stage renal

disease: comparison and stability across treatment modalities," *Health and Pysychology,* Vol. 9, No. 2, pp. 117–42. (Medline).
Dumit, Joe, et al. (1992) "Cyborg Anthropology," unpublished ms.
Dutton, Diana B. (1988) *Worse Than the Disease: Pitfalls of Medical Progress,* (Cambridge, UK: Cambridge University Press).
Foldy, Seth (1978) "A Tale of Two Cultures," unpublished ms.
———, (1979) "Interview with Dr. Yuki Nosé—June 20," unpublished ms.
———, (1990) "Metaphor as Illness: The Underclass Concept and Medical Care," presented at the 118th Annual Meeting of the American Public Health Association, Oct. 3.
Foss, Laurence, and Kenneth Rothenberg (1987) *The Second Medical Revolution: From Biomedicine to Infomedicine,* Boston: New Science Library/Shambhala Publications.
Fox, Renée C., and Judith P. Swazey (1974) *The Courage to Fail,* Chicago: University of Chicago Press.
Friedman, E. A. (1987a) "ISAO Proffers a Marvelous Cover for Acting Out Fantasies," *Artificial Organs,* Vol. 11, No. 3, p. 193.
———, (1987b) "John P. Merrill: America's First Nephrologist," *Artificial Organs,* Vol. 11, No. 6, p. 437.
Funck-Bretano Jean-Louis (1986) "John P. Merrill Memorial Presidential Address," *Artificial Organs,* Vol. 10, No. 3, pp. 207–10.
Gaffney, F. Andrew, and Barry J. Fenton (1984) "Barney B. Clark, DDS: A View From the Medical Services," *Archives of General Psychiatry,* Vol. 41, Sept., pp. 917–8.
Galletti, Pierre M. (1986) "Artificial Organs Science: Three Spirits Revisited," *Artificial Organs,* Vol. 10, No. 5, pp. 385–6.
Gee, William F. (1975) "A History of Surgical Treatments of Impotence," *Urology,* Vol. 5, No. 3, pp. 401–5.
Gee, William F., J. William McRoberts, James O. Raney, and Julian S. Ansell (1974) "The Impotent Patient: Surgical Treatment with Penile Prosthesis and Psychiatric Evaluation," *Journal of Urology,* Vol. 111, Jan., pp. 41–3.
Gerhardt, H. J., and H. Wagner (1987) "Abstract: The Berlin Cochlear Implant—First Experiences," *Artificial Organs,* Vol. 11, No. 4, p. 342.
Gonsalves-Ebrahim, L., and M. Kotz (1987) "Abstract: The physchological impact of ambulatory peritoneal dialysis on adults and children," *Psychiatric Medicine,* Vol. 5, No. 3, pp. 177–85. (Medline).
Graupe, Daniel, and Kate H. Kohn (1994) *Functional Electrical Stimulation for Ambulation By Paraplegics,* Malabar, Florida: Krieger Publishing.

Gray, Chris (1989) "The Cyborg Soldier" in Levidow and Robins, eds. *Cyborg Worlds: Programming the military information society,* London: Free Association Press; 1992, New York: Columbia University Press, pp. 43–73.

———, (1993) "The Culture of War Cyborgs: Technoscience, Gender, and Postmodern War," *Research in Philosophy & Technology,* special issue on gender and technology, Joan Rothschild, ed., in press.

———, with Heidi J. Figueroa-Sarriera and Steven Mentor (1996) *The Cyborg Handbook,* New York: Routledge.

Handelsman, Harry (1990) "Implantation of the Automatic Cardioverter-Defribrillator: Noninducibility of Ventricular Tachyarrhythmia as a Patient Selection Criterion," *AHCPR Health Technology Assessment Report No. 10,* U.S. Dept. of Health and Human Services.

Haraway, Donna (1985) "A Manifesto for Cyborgs: Science, Technology, and Socialist Feminism for the 1980s," *Socialist Review,* No. 80, pp. 65–107.

———, (1990a) *Simians, Cyborgs and Women: The Reinvention of Nature,* New York: Routledge Keegan Paul.

———, (1990b) "Cyborgs at Large: Interview with Donna Haraway," (by Constance Penley and Andrew Ross) in Penley and Ross, eds. *Technoculture,* Minneapolis: University of Minnesota Press, pp. 1–20.

———, (1990c) "The Actors are Cyborg, Nature is Coyote, and the Geography Is Elsewhere: Postscript to 'Cyborgs at Large,' " in Penley and Ross, eds. *Technoculture,* Minneapolis: University of Minnesota Press, pp. 21–6.

Harken, Dwight E. (1991) "Pacemakers, Past-makers and the Paced: An informal history from A–Z (Aldini to Zoll)," *Biomedical Instrumentation and Technology,* Vol. 25, No. 4, pp. 299–321.

Hartouni, Valerie (1990) "Containing Women: Reproductive Discourse in the 1980s," in Penley and Ross, eds. *Technoculture,* Minneapolis: University of Minnesota Press, pp. 27–56.

Hilton, Bruce (1992) "How to tell when someone is dead," *Cleveland Plain Dealer,* July 19, p. 3-C.

Hoffman, Allan S. (1991) "Abstract: Future Trends in Biomaterials Research and Development," *Artificial Organs,* Vol. 15, No. 4, p. 302.

Hogle, Linda (1996) "Tales from the Cryptic: Technology Meets Organism in the Living Cadaver," in Gray, ed. *The Cyborg Handbook,* New York: Routledge, pp. 203-18.

Hori, Motokazu (1986) "Artificial Liver: Present and Future," *Artificial Organs,* Vol. 10, No. 3, pp. 211–3.

Hudak, Stephen (1992) "Boy keeps charging on artificial leg," *Cleveland Plain Dealer,* June 29, p. 4-B.

Johnson, Kristen E., Mikel Prieto, Lyle D. Joyce, Marc R. Pritzker, Charles R. Jorgensen, and Robert W. Emery (1991) "Abstract: World Experience with Total Artificial Heart Implantation: A Registry Report," *Artificial Organs,* Vol. 15, No. 4, p. 279.

Kantrowitz, Adrian (1983) "Threshold," *Artificial Organs,* Vol. 7, No. 2, p. 149.

Kaufert, Joseph M., and David Locker (1990) "Rehabilitation Ideology and Respiratory Support Technology," *Social Sciences and Medicine,* Vol. 30, No. 8, pp. 867–77.

Kimbrell, Andrew (1992) "Body Wars: Can the human body survive the age of technology?" *Utne Reader,* May-June, pp. 52–64.

Kjellstrand, C. M., F. Ericsson, A. Traneous, L. O. Noree, and L. E. Lins (1989) "Abstract: The wish for renal transplantation," *ASAIO Transactions,* Vol. 35, No. 3, July-Sept., pp. 619–21. (Medline).

Kline, Nathan S., and Manfred Clynes (1961) "Drugs, Space, and Cybernetics: Evolution to Cyborgs," in Bernard E. Flaherty, ed. *Psychophysiological Aspects of Space Flight,* New York: Columbia University Press, pp. 345–71.

Kolff, Willem J. (1966) "Medical Travelogue: About Artificial Organs, Kidney Transplantation, and Unrelated Medical Experiences in Europe, Fall 1964," *Ohio State Medical Journal,* Vol. 62, No. 4, April, pp. 323–8.

———, (1977) "Exponential Growth and Future of Artificial Organs," *Artificial Organs,* Vol. 1, No. 1, pp. 8–17.

———, (1979a) "The History of the Artificial Kidney," *Transactions of the American Society for Artificial Internal Organs, 25th Anniversary Memorabilia Edition,* Vol. 1, (June 5, 1955), pp. 1–19.

———, (1979b) "Questions and Predictions," in Felix Unger, ed. *Assisted Circulation,* Berlin: Springer-Verlag, pp. 11–2.

———, (1989) "The Artificial Heart, the Inevitable Development: Will it be in the U.S. or abroad?" *Artificial Organs,* Vol. 13, No. 3, pp. 183–4.

Kolff, W. J., and John Lawson (1979) "Spare Parts for People," *Encyclopaedia Britannica Yearbook of Science and the Future,* pp. 42–55.

Levidow, Les, and Kevin Robins, eds. (1989) *Cyborg Worlds,* London: Free Association Press.

Lindgren, Nilo (1965) "The Artificial Heart—Exemplar of medical-engineering enterprise," *IEEE Spectrum,* Sept., pp. 67–83.

Lubeck, Deborah P., and John P. Bunker (1982) *Case Study #9, The Artificial Heart: Cost, Risks, and Benefits,* Washington, D.C.: Congress of the United States Office of Technology Assessment.

McKenzie, D. S. (1967) "Still a Long Way to Go," *Artificial Limbs,* Vol. 11, No. 2, Autumn, pp. 1–4.

Mayer, V., M. Frey, J. Holle, M. Kern, H. Stohr, G. Schwanda, and H. Thoma (1987) "Abstract: Results of a Clinical Study with Paraplegic Patients Using an Implanted System for Functional Neurostimulation of Their Lower Extremities," *Artificial Organs,* Vol. 11, No. 4, p. 339.

Morris, P. L., and B. Jones (1989) "Abstract: Life satisfaction across treatment methods for patients with end-stage renal failure," *Medical Journal of Australia,* Vol. 150, No. 8, April 17, pp. 428–32. (Medline).

Murray, Joseph E. (1992) "Human Organ Transplantation: Background and Consequences," *Science,* Vol. 256, June 5, pp. 1411–6.

Nosé, Yukihiko (1985) "Therapeutic Artificial Organs: Future Perspectives," *Artificial Organs,* Vol. 9, No. 1, pp. 7–11.

———, (1986) "Totally Implantable Artificial Organ: Cardiac Prosthesis," *Artificial Organs,* Vol. 10, No. 2, pp. 102–13.

———, ed. (1990) "Abstracts from the Third Vienna International Workshops on Functional Electrostimulation, Basics, Technology, and Clinical Applications," *Artificial Organs,* Vol. 14, No. 6, pp. 473–88.

Petrie, K. (1989) "Abstract: Psychological well-being and psychiatric disturbance in dialysis and renal transplant patients," *British Journal of Medical Psychology,* Vol. 62, pt. 1, pp. 91–6. (Medline).

Preston, Thomas (1982) "Appendix A: The Artificial Cardiac Pacemaker" in *The Implications of Cost-Effectiveness Analysis of Medical Technology—Case Study #9: The Artificial Heart: Cost Risks, and Benefits,* Washington, D.C.: Congress of the United States Office of Technology Assessment, pp. 41–3.

Putti, V. (1929) "Historic Artificial Limbs," *American Journal of Surgery,* Vol. VI, No. 1, Jan., pp. 111–7; No. 2, Feb., pp. 246–53.

Renshaw, Domeena (1979) "Inflatable Penile Prosthesis," *Journal of the American Medical Association,* Vol. 241, No. 24, pp. 2637–8.

Reswick, James B., and Lojze Vodovnik (1967) "External Power in Prosthesis and Orthotics, an Overview," *Artificial Limbs,* Vol 11, No. 2, Autumn, pp. 5–21.

Romm, Sharon (1988) "Arms by Design: From Antiquity to the Renaissance," *Plastic and Reconstructive Surgery,* Vol. 84, No. 1, July, pp. 158–63.

Scott, F. Brantley, Gary J. Byrd, Ismet Karacan, Peter Olson, Larry E. Beutler, and Samual L. Attia (1979) "Erectile Impotence Treated with an Implantable, Inflatable, Prosthesis: Five Years of Clinical Experience," *Journal of the American Medical Association,* Vol. 241, No. 24, pp. 2609–12.

Silber, Jerome, and Sydelle Silverman (1958) "Studies in the Upper-Extremity Amputee: VII. Psychological Factors," *Artificial Limbs,* Vol. 5, No. 2, pp. 88–116.

Skurzynski, Gloria (1978) *Bionic Parts for People: The Real Story of Artificial Organs and Replacement Parts,* New York: Four Winds Press; illus. by Frank Schwarz.

Smith, Harmon L. (1990) "Book Review: *Who Lives? Who Dies? Ethical Criteria in Patient Selection* by John Kilner, New Haven: Yale University Press, 1990" in *Social Sciences and Medicine,* Vol. 31, No. 10, pp. 1187–9.

Stiglebrunner, H. K., I. J. Hochmaier-Desoyer, E. S. Hochmaier, E. L. V. Wallenberg, and K. Burian (1987) "Abstract: The Vienna Cochlear Implant Program," *Artificial Organs,* Vol. 11, No. 5, p. 430.

Strauss, Michael J. (1984) "The Political History of the Artificial Heart," *New England Journal of Medicine,* Feb. 2, pp. 332–6.

Tiefer, Lenore, Beth Pederson, and Arnold Melman (1988) "Psychological Follow-up of Penile Prosthesis Implant Patients and Partners," *Journal of Sex and Marriage Therapy,* Vol. 14, No. 3, Fall, pp. 184–201.

Tiefer, Lenore, Steven Moss, and Arnold Melman (1991) "Follow-up of Patients and Partners Experiencing Penile Prosthesis Malfunction and Corrective Surgery," *Journal of Sex and Marriage Therapy,* Vol. 17, No. 2, pp. 113–27.

Tzamaloukas, A. H., P. G. Zager, B. J. Quintana, M. Nevarez, K. Robers, and G. H. Murata (1990) "Abstract: Mechanical cardiopulmonary resuscitation choice of patients on chronic peritoneal dialysis," *Peritoneal Dialysis International,* Vol. 10, No. 4, pp. 299–302. (Medline).

Van Alsté, J. A. and P. H. Veltink (1987) "Abstract: Sensitivity and Selectivity of Nerve Stimulation with Karsussell-Electrode," *Artificial Organs,* Vol. 11, No. 5, p. 430.

Van Citters, Robert L., Catherine B. Bauer, Lois K. Christopherson, Robert C. Eberheart, David M. Eddy, Robert L. Frye, Albert R. Jonsen, Kenneth H. Keller, Robert J. Levine, Dwight C. McGoon, Stephen G. Pauker, Charles E. Rackley, Vallee L. Willman, and Peter L. Frommer (1985) "Artificial Heart and Assist Devices: Directions, Needs, Costs, Societal and Ethical Issues,"

Artificial Organs, Vol. 9, No. 4, pp. 375–415; reprinted from NIH publication 85-2723.

Van Nieuwkerk, C. M., R. T. Krediet, and L. Ariez (1990) "Abstract: Voluntary discontinuation of dialysis treatment by chronic dialysis patients," *Ned-Tijdschr-Geneeskd,* Vol. 134, No. 32, August 11, pp. 1549–52. (In Dutch, translated abstract on Medline).

White, Robert J. (1992) "Interview" at the Allen Memorial Library, Cleveland, Ohio, July 21. Dr. White is the director of neurosurgery at MetroHealth Medical Center and professor of surgery at Case Western Reserve University.

Wolcott, D. L., A. R. Nissenson, and J. Landsverk (1988) "Abstract: The Quality of life in chronic dialysis patients. Factors unrelated to dialysis modality," *General Hospital Psychiatry,* Vol. 10, No. 4, July, pp. 267–77. (Medline).

Zimmermann, E. (1989) "Abstract: Quality of life in artificial kidney therapy," *Wien-Klin-Wochenschr,* Vol. 101, No. 22, Nov. 24, pp. 780–4. (In German, translated by Medline).

8
SOME BODY FANTASIES IN CYBERSPACE TEXTS: A VIEW FROM ITS EXCLUSIONS[1]
Heidi J. Figueroa-Sarriera

> All technique is technique of the body. It represents and amplifies the metaphysical structure of the flesh.
>
> —Merlaeu-Ponty

I shall begin with a basic statement: high technology discourses of the present have restored time and space categories to their (only recently lost) privileged position within techno-cultural discourse and practices. Since the Renaissance these categories, time and space, have been launched from yet another one, the body. And the body is articulated through technology, which continually changes over time. As we will see, the body is centered as an axis that rearticulates different high-tech social myths, which the history of technology can help us interpret. Through history, the body variously, and sometimes simultaneously, has been seen as a volume in disintegration, as a space of replacement or displacement, and as disposable stuff, among other possibilities. Studying contemporary technology in historical perspective is only part of understanding high-tech social myths. They must be interpreted as well.

My basic working premise is that all social discourse—including scientific discourse—is necessarily rhetoric. Like Habermas and many others, I believe that the bond between rhetoric and argument is already consolidated; scientific discourse included. This rational discourse par excellence happens to be one of the most fertile fields of rhetorical strategies. This is especially true of the focus of this paper: those rhetorical figures which are prominent in texts on what we know today as "cyberspace" (cybernetic space), a term originally coined by William Gibson (1984) in his science fiction novel *Neuromancer*. This technology, together with nanotechnology (which is the fusion of biological, microelectronic, and micromechanical techniques) and genetic engineering, is one of the most promising and seductive of the contemporary technoscientific fields.

In order to accomplish my purpose, analyze rhetorical-historical strategies as power effects on certain texts regarding cyberspace, I have included three interrelated reflections. First, there is a brief review of the ways in which conceptions of time and space have changed from the Renaissance to the present. Second, there is a reflection on the centrality of the body as a space of semantic negotiation of new signifiers anchored in two time-space versions. And third, there is an analysis of some common metaphors in cyberspace texts, trying to suggest the ways in which these discourses are articulated as power effects in two ways: constructing asymmetrical social relations by excluding certain populations and at the same time reinvigorating the desire for social equality and liberatory practices. Finally, I wish to share some thoughts on the imperative recycling of the psychological discipline, since high technology devices are organizing social life forms as well as their significances.

CYBERSPACE

Cyberspace—as with any category which appeals to and summons our deepest passions—proliferates in an impressive intertextuality. William Gibson defined it as follows:

> Cyberspace. A consensual hallucination experienced daily by billions of legitimate operators, in every nation, by children being taught mathematical concepts ... A graphic representation of data abstracted from the banks of every computer in the human system. Unthinkable complexity. Lines of light ranged in the nonspace of the mind, clusters and constellations of data. Like city lights, receding ... (Gibson, 1984, p. 51)

In Gibson's novel there are digital cowboys who directly connect to the Matrix, the computational infrastructure which produces this new domain.

In general terms, we could argue that the concept refers to any electrodigitalized space provided by the encounter of several participants in a virtual scenario. For instance, there was VIDEOPLACE, developed during the last five years of the 1970s by Myron Kureger and mainly dedicated to games and art. Then there are the virtual reality (VR) devices developed by NASA and rendered popular during the eighties. These consist of gloves and a small screen covering both eyes. This technology has inspired scientists to develop programs capable of creating 3D molecular designs and remote control robots which transmit certain sensory experiences to the operator.

Likewise, the press and the so-called popular literature have anticipated new sexual ("teledildonics") and electronic/psychedelic modalities. Suddenly, cyberspace has become a huge cultural metaphor encompassing a whole series of computer networks, fax lines, virtual showrooms capable of revealing the architectronic features of a building yet-to-be-built including its interior furnishings, virtual museums, and collective graphics—all in permanent movement. This technology is also used by the military as in the simulations and computer displays that played such a large role in the Persian Gulf War.

In order to put this discussion in historical perspective I shall begin with the four principles which, according to David Harvey (1990), dominated conceptions of time and space over the last four centuries since the Renaissance. First of all, we have the fixed point-of-view, and ethnocentric point-of-view of the discovery voyages which revealed a larger world to be absorbed, represented, and of course, colonized. Consequently, geography becomes a value of use and exchange. Then, we have "representationalism," in which the individual and his/her visual abilities appear to be capable of representing what he/she sees as real, thus generating the aura of the artist as a creative genius-entrepeneur. Accumulation of wealth, power, and capital have become linked to knowledge's personalization; the individual appears to be dominating physical space while at the same time dominating time, its representation. For this purpose the fundamental principles of perspective are developed, objectivity thus assuming a positive value in said "representationism." Finally, rationality is refocused in measurement, control, and centralization tasks of properties and the articulation of new power-knowledges.

Rearticulating space as a "fact" of nature means that the rational ordering of space will become the integrating condition of human emanicipation. Space and time must be organized not to reflect the glory of God, but to celebrate and facilitate the liberation of "Man" as a free and responsible individual, in accordance with the parameters of the "self" in Modernity. Dominion over nature favored scientific prediction, through social engineering, rational planning, and the institutionalization of homogeneous time; space was formed by the limits of thought and action. Realistic narrative, which presupposes a coherent narration, describing one event after another through lineal time, gained popularity.

In the nineteenth century, and at the beginning of the twentieth, there was a second Modernist revolution which put realistic narrative and objectivism into crisis, and the Euclidian perspectivist space seemed to disappear (Lefebre, 1991). The world is reorganized, not in a Newtonian but in an Einsteinian universe. Categories of time and space become relative and compressed, speaking in terms of time-space, thus inaugurating the "dash" culture as a graphic vehicle to express the lack of conformity with dichotomic polarities of Occidental thought. According to Einstein's general theory, the universe is not a rigid whole in which an independent matter is situated in space and time equally independently, instead it is a plastic, variable continuum, subject to constant change: the stars modify the space in which they move.

Harvey claims that there are two versions of this time-space union: the celebration of the annihilation of space through time (especially with technological progress in communications media, which allow the immediate connection of two or more distant events in the same space) and the reorganization of labor and non-labor through time spatialization (in other words, time control through space organization, according to Fordism and Post-Fordism principles, for instance). Assuming these two versions I would add that in the celebration of the annihilation of space through time the body appears as an active consumer-producer of contradictory social images. Meanwhile, through time spatialization the body appears as a rebel product of disciplinary regimes.

Recently, several authors have dedicated many pages to the cultural consequences of this time-space compression, seen as the disappearance of time and space as material dimensions of social life (Virilio, 1984, 1988, is an outstanding example). However, when we approach texts of high technology, especially those focusing on philosophical and practical problems of cyberspace, I believe that it is more of a matter of the cultural process of displacement and replacement of said categories, than of disappearance. Within these cultural processes, the body, as a physical, organic, and material body, plays a fundamental role as a paradoxical point of departure. The technologized body becomes the sign and symptom of the contradictions that capital accumulation forms have had historically with regard to the human body as an object of production and consumption. On the one hand, the body can subordinate itself to the functionalism and pragmatism of accumulation forms and their disciplinary regimes, dedicated to safeguard social order and control; and on the other hand, simultaneously, this same body deceives its guardians, it resists and sabotages them. This game filters into the forms in which engineers, programmers, and specialists of all kinds, talk about their present or potential creations. These forms are promulgated through metaphors, as I hope to show.

METAPHORS OF CYBERSPACE

Let us consider some examples focusing on three of the metaphors most often enunciated in the examined texts: the magical, the organic, and the nautical.

Instances of the magical metaphor with its particular power effects and social implications, abound. Nicole Stenger (1992), a computer animation artist who in 1989 was working on a virtual world project for Washington University, argues, "cyberspace is like Oz—it is, we get there, but it has no location" (p. 53). In her text, she goes on to establish a paradigmatic chain of signifiers between magic (sacredness) as an unavoidable dimension of consciousness, and play as an essential condition of life in society. Cyberspace emerges as a space to celebrate and a place to let the imagination run free. The computer itself appears as a vehicle which positively transforms the capacities of sensory organs.

> Working with computers changes our sensitivity to light, to depth, makes our dreams more vivid, facilities the use of metaphors in language . . . This is why we can expect the speed of 'real time' to help us project into cyberspace some of our dearest phantasms, some of our worst monsters. This power of revelation and embodiment will be felt by many to be the utmost obscenity. Let's not forget that both Hitler and Stalin are known for having banned the publication of fairy tales. Freedom of imagination is feared by most powers. (p. 57)

She ends her work by proclaiming,

> According to Sarte, the atomic bomb was what humanity had found to commit collective suicide. It seems, by contrast, that cyberspace, though born of a war technology, opens up a space for collective restoration, and for peace. As screens are dissolving, our future can only take on a luminous dimension! Welcome to the New World! (p. 58)

We can establish certain contrasts between this form of textual presentation and the forms of presentation of the popular and expert discourse regarding electricity in the nineteenth century. Carolyn Marvin (1988) tells us that in popular science the body appears as a point of reference from which the world is understood, while the culture of scientific experts attaches to the disembodied reasoning of formal theories. However, when specialists projected themselves into the popular culture, they appealed to religion and magic as a fundamental rhetorical vehicle, since these were the very expression forms of oral culture. Another characteristic of this discourse on electricity was that the experts resisted the body as a source of pleasure since, publicly, popular attitudes were suspicious about pleasurable uses of the human body.

Discourses on cyberspace, both within the scientific community as well as when they are projected to the public, appeal to the magical-religious qualities of the devices. However, and in contrast, they reinscribe the sensoral body as a privileged and legitimate source of pleasure, sometimes with sophisticated stylistic subterfuges and at other times, with more evident forms, the reinsertion of the body as a pivot of reflection is recognized. Stenger also says,

> Descartes would say that 'the senses cheat us.' Now the senses are back as the only reasonable means of information when the acceleration of modern warfare makes the time needed for abstract analysis obsolete. (p. 51)

However, they do not refer to the body as we know it ordinarily, the material body appears as a space displaced by another one, fundamentally built with "bits" and "bytes." This new body image assumes a completely protean quality which reinstalls us within the aesthetic value of not only the ephermeral but also of "virtualness," of that which does not belong to conventional forms.

Meanwhile, rhetorical strategies reveal themselves as power strategies. They appeal to our desires, our ghosts, our fears and dearest longings with the promise of the "discovery" of a better world. Yet, the knowledge allowed by the construction of such an esoteric world appears to be obliterated. This knowledge is reserved for the god-experts who set forth the precepts of the sacred world—precepts that are debated and established through the arbitrariness of capital accumulation, in which marketable and militarily oriented projects have a greater future. In this world, users (now called participants) participate only partially. But promises have always entailed a terrible risk; the "other" can begin to believe them. Hence, the paradoxical knot and the refutable potential of social control projects: if we assume this new world, the claim of a "cybercitizen" with the ability to participate at all levels should soon be expected. This is a problem which has already been anticipated by some designers and which brings out the secret's paradoxical permanence—through passwords and other forms of access control in computer systems—in the alleged free flow of the information world. In fact, the cybercitizen who juggles, the saboteur, the one who, while consuming these devices, produces results unexpected by the designers—such as ecological networks with impossible electrodigitalized classes—already lives amongst us.

The body's displacement is accompanied by a sensory reappraisal as a resource of knowledge, especially in areas where such an experience had not yet been possible. For instance, this reappraisal has been molded in devices designed to "feel" the molecular union and breakage in chemical reactions through a virtual reality. It is also important to indicate that in other texts, this reappraisal is accompanied by a devaluation of the body's organic substrate thus causing the articulation of an image of same as scatological and disposable object. However, once again, the articulation of said discourse is not univocal, since it also suggests the advantages provided by the new technology safeguarding the biological body from the dangers implied in long voyages to desolate places, as well as the accomplishment of high risk activities.[2]

In debates on design, the question of the codes of these new bodies with personified appearances becomes very relevant. Experts believe that this is one of the clues for the success of said systems. A key principle applied to cyberspace design is precisely visibility. To constitute a space within space, a volume that is visible to the "other," is an essential requirement. The actual physical body is (most of the time) a drawback, loaded with limitations, overwhelmed by suspicious technical nuances and features, tired and pathologically mutant, and it could be temporarily substituted by another or others.

EMBODIMENTS OF CYBERSPACE

Restructuring urban space therefore implies a redefinition of reconstructed volumes in what has been called "hyper-reality." It also implies a greater communication capacity together with a displacement of the body through access and representation codes within the electrodigitalized communication network. This scenario suggests to psychology the need to redefine the so-called interpersonal communication processes (among which face-to-face interaction has enjoyed a privileged position) to approach the idiosyncrasies and dynamisms of these new forms of social relations.

I propose that psychologists could consider, among other possible agendas:

1. Examining the textual analysis of narrative forms from which the text and its forms of social relations are organized and disorganized.
2. Exploring the study of forms of narrative components corresponding to the scientific-technical work in these projects, filtering them through the rearticulation of discursive practices of post-anthropological daily life, following Foucault's intuitions about the end of the anthropological period of philosophy and Western culture. "It is no longer possible to think in our day other than in the void left by man's disappearance. For this void does not create a deficiency; it does not constitute a lacuna that must be filled. It is nothing more, and nothing less, than the unfounding of a space in which it is once more possible to think" (Foucault, 1973, p. 325).
3. Studying the new subjectivities which arise from these new social understandings, such as the work already published about the "fragmented self" or the "decentered self" (for instance Lash and Urry, 1987; Figueroa-Sarriera and López, 1993, among others) or the "saturated self" (Gergen, 1992).
4. Mapping the forms of electrodigitalized restructuring of new social actors which are territorialized and deterritorialized within the accelerated flow of post-industrialized or quasi-post-industrialized societies, that live under pressure with the different hegemonic projects.[3]
5. Using the social and historic forms of the construction of the world of objects ("cyberobjects" which can either be present or not in cyberspace), their relations, and the subject's paradoxical route into this "new world."

This last approach could even imply a reconsideration of psychoanalytic study,[4] particularly with regard to those aspects which emphasize the analysis of object relations within the process of subjectivity formation, the Lacanian reformulations regarding the constitution of the object and the Thing, or in its most extreme version, the Theory of Object Relations in Melanie Klein's school.

Alan Wexelblat (1992), a staff systems engineer with Bull Worldwide Information Systems, is working on the underlying mechanisms of cyberspace and he is trying to translate those into a commercial product for Bull. He has invented a concept he calls "semantic space" in order to organize a type of technology for users who must deal with very abstract information. When someone wants to view information unrelated to physical location and has to deal with the problem of how to lay out the information in a consistent and understandable manner, Wexelblat's insight is supposed to help. One example would be if you want to view a complex system of software modules without external reference to layout. He defines "semantic space" as a "general mechanism that can be used to help solve problems of placement and composition" (p. 256). I shall not entertain you with the details of this proposal but I will extract a narrative segment with regard to his vision of objects and space in cyberspace.

> In ordinary computer systems movement does not necessarily have the same societally constructed connotations. In most such systems, movement is merely a means of stopping interaction with one object and beginning interaction with another. Movement is necessary because objects are at some distance from one another, but the act of movement itself has no meaning.
> ... However, this loss of meaning need not happen in a semantic cyberspace. An effect of constructing cyberspace along semantic dimensions is to render the actions of motion meaningful in and of themselves ... the space itself has meaning, possible even when no objects are present ...
> ... This allows us to perform visual searches quickly and helps us find objects that may more or less closely match our desires ... (p. 265).

First of all, Wexelblat centers his attention on the object, its relations, and the movement it directs towards the subject, from one to the other. In other words, the ordering of space in movement becomes significant in the sense that Michel de Certeau (1987) talks about of "spatial practices" (ways of doing) in a transhuman city where the motions of walking are spatial creations and where the first definition of walking, then, would be a space of uttering. De Certeau claims that walkers transform every spatial signifier into something else, creating discontinuity by choosing among the signifiers of the spatial scene or by altering them through the use he/she makes of them. Thus, the prevailing metaphors are oriented towards "doing something" rather than "carrying something." In semantic space the meaning of objects is changed by moving them along dimensions.

Secondly, we presuppose paradoxical processes within cyberspace, which imply identification/separation of the subject with regard to the fluid object of desire and which advance the image of the "cyborg" beyond the limits of science fiction, as a political myth of the end of the millennium, as suggested by Donna Haraway (1992).

It is important to point out that cultural analysis has already recognized the commonality between engineers, workers, and sexual entrepreneurs: they are all experts in designing tokens that are easily recognized as objects of desire. The erotic component (even though it may still be pre-orgasmic) is definitely present in debates regarding these systems.

The notion of "Cyborg envy" has even been proposed, about which Allucquere Rosanne Stone (1992) says:

> To enter the discursive space of the program is to enter the space of a set of variables and operators which the programmer assigns names. To enact naming is simultaneously to possess the power of and to render harmless, the complex of desire and fear that charge the signifiers in such a discourse; to enact naming within the highly charged world of surfaces that is cyberspace is to appropriate the surfaces, to incorporate the surfaces into one's own. Penetration translates into envelopment. In other words, to enter cyberspace is to physically put on cyberspace. To become a cyborg, to put on the seductive and dangerous cybernetic space like a garment, is to put on the female. (p. 109)

Culturally, the reinvention of the body in the terms I have been discussing has produced several reactions. Among these we can point out the celebratory movement and the emancipating and subversive potential of these devices and of the promises they call upon. We also have to reappraise the naturalistic body, which has manifestations that are as diverse as vegetarianism, physiculturalism, and the zeal to have an ecological, unpolluted body.

Notwithstanding the discursive displacement of the biological body, I must also add the metaphoric image where the organic appears as one of the fundamental structuring axes of this discourse. The sophisticated mechanism which allows communication in cyberspace, entries and exits, is called "matrix." The origin of this word is Latin and means "mother." Different studies have suggested the perseverance of masculine fantasies in technological designs, among which are the type of procreation desire which has been called "uterus envy." Masculine inability to conceive inside the body seems to sometimes summon the desire to conceive outside of it, and this is done through technological (often warlike) devices. Let us not forget that the first uranium bomb used, on Hiroshima, was called by its direct creators "Little Boy" (Easlea, 1983). If we assume "cyberspace" as a discursive field, we then have the possibility of identifying multiple subject positions within it.

Yet, cyberspace is also evoked as a "habitat," referring once more to that which is organic. The initial point of reference of this notion is the fauna and flora forms of life. Thus, "cyberspace" is molded as a new space in nature. As a matter of fact, Habitat is the name of one of these systems; this one designed by Chip Morningstar and Randall Farmer, as a large-scale social experiment through the Tymnet network. Likewise, the image of "symbiosis" to describe the relation between the devices which allow the experience of cyberspace, belongs to the biological discourse. Suffice it to say that this metaphor not only suggests a certain naturalness in technological processes and devices, but it also serves the purpose of closing distinctions between that which belongs to nature and that which belongs to technology; or what we might call the human-machine compression on the social organizational level.

It is interesting to add that these rhetorical resources, particularly those which appeal to erotism and seduction, are almost invariably accompanied by dominating and/or aggressive images. There are numerous cases.

For example, Michael Benedikt, architect and consultant on software design, indicates that in cyberspace:

> Because virtual worlds—of which cyberspace will be one—are not real in the material sense, many of the axioms of topology and geometry so compellingly observed to be an integral part of nature can there be violated or reinvented, as can many of the laws of physics. (Benedikt, 1992b, p. 119).

Facing the question of "What is reality?" he solemnly replies "Reality is death" (Benedikt, 1992a, p. 14).

Michael Heim, a well-known consultant to the computer industry, describes the erotic ontology of cyberspace.

> Our love affair with computers, computer graphics, and computer networks runs deeper than aesthetic fascination and deeper than the play of the senses. We are searching for a home for the mind and heart. Our fascination with computers is more erotic than sensuous, more deeply spiritual than utilitarian ...
>
> ... The computer's allure is more than utilitarian or aesthetic; it is erotic. Instead of a refreshing play with surfaces, as with toys or amusements, our affair with information machines announces a symbiotic relationship and ultimately a mental marriage to technology ...
>
> ... The world rendered as pure information not only fascinated our eyes and minds, it captures our hearts. We feel augmented and empowered. Our hearts beat in the machines. This is Eros (Heim, 1992, p. 61).

NAVIGATING CYBERSPACE'S MEANINGS

Finally, the nautical metaphor, which we also saw in the conception of semantic space previously discussed, stands out in these texts. The navigation metaphor assumes multiple forms: participants are "cybernauts," the entry process to this space is called "immersion," the goal is to reach a "new world," the process of displacement itself within cyberspace is called "navigation." These statements reflect echoes of past and present feats of annihilation, extermination, and genocide in the history of our people while echoes of mastery, wealth, and life in other countries' official histories. Like the shadow of Peter Pan, the indomitable resentment always returns to spoil the promise of adventure.

It seems that within the context of asymmetric social relations, the struggle to control space (be it conventional or electrodigitalized) works as an indisputable source of attempts at systematic hegemony. Latin American countries—for better or for worse—resist their subordinated and peripheral role in economic globalization processes. They constantly demand more participation, and equal economic and political treatment. This places Latin America in transit towards production and technological appropriation forms, as shown in a festive and superlative form, by the Carnival del Rio of 1993. Participants selected high technology motifs as a form of fighting conservative censorship against naked bodies, and at the same time, revealing in a manifest way, the technocratic politics of elites. The Carnival turned into a show of poetic vengeance. The technologized voluptuous body dances as a contesting political practice which overflows the conventional forms of politics.

This element of passionate technological fascination has also been emphasized in the writings of experts. Marcos Novak, a researcher in the area of algorithmic composition, cyberspace, and the relationship between architecture and music, referred to cyberspace fascination:

> The root of this fascination is the promise of control over the world by the power of the will. In other words, it is the ancient dream of magic that finally nears awakening into some kind of reality. But since it is technology that promises to deliver this dream, the question of 'how' must be confronted. Simply stated, the question is, What is the technology of magic? . . .
> . . . Cyberspace is poetry inhabited, and to navigate through it is to become a leaf on the wind of a dream. (Novak, 1992, pp. 228–9)

This affirmation has the merit of recognizing the legacy of alchemy—and even more so, of that ancient mythology which always venerated movement as a sign of life and a symbol of deity—in the development of present technology. But it is also an acknowledgement of the role played by aesthetics, passion, and fantasy in the production-consumption of rational devices.

> If every instrument could accomplish its own work, obeying or anticipating the will of others . . . If the shuttle could weave, and the pick touch the lyre, without a hand to guide them, chief workmen would not need servants, nor masters' slaves. (Quoted in Malone, 1978, p. 7).

Nevertheless, at this moment it is crucial to state a question, which maybe should be posed before Novak's, and it is: What causes the magic of technology? And maybe a sneaky answer would be: *the magic of technology is precisely the promise of a technology of magic.*

Then other questions and silences follow: Who promises? What for? To whom? Where from? To assume that they come from Latin America and the Caribbean implies the further assumption of the uncertain universe of words in a world where—as salsa and merengue composer and singer, Juan Luis Guerra from the Dominican Republic says—"los vivos son sobrevivientes" (those who are alive are survivors).

NOTES

1. This paper was presented at the International Society for Theoretical Psychology Conference, April 25–30, 1993, at Saclas, France. It was sponsored by the National Endowment for the Humanities Summer Seminar Program. The Summer Seminar on "Technology and American Culture," directed by Dr. Carroll Pursell of Case Western Reserve University, was the intellectual context from which this paper was born.

2. Figueroa-Sarriera Heidi, (1981) *Metaforas de Persona en Textos Sobre Inteligencia Artificial y Robotica.* Doctoral Dissertation, Dept. of Psychology, Recinto de Río Piedras, University of Puerto Rico; and "Children of the Mind with Disposable Bodies," in Chris Hables Gray, Heidi J. Figueroa-Sarriera, and Steven Mentor, eds., *The Cyborg Handbook* (New York: Routledge, 1996)

3. For instance, the electrodigitalized reordering of banking in Puerto Rico constantly and dramatically recreates the population with its fluid social-economic identities as credit or cash producers-consumers. The level of participation in the big spectacle of present or potential, real or imaginary consumption, in a country where a

high level of poverty and illiteracy has publicly been recognized is in contradiction to the annexationist project which pretends to turn Puerto Rico into state number 51 of the United States.

4. Sherry Turkle (1990) has suggested that artificial intelligence (AI) projects—especially those dealing with "emergent" concepts—raise the possibility that psychoanalysis can be reinvigorated based on the uncertainties debated through AI, and with a greater impact on everyday life culture. Psychological cultures do not exist only in the world of professionals. Artificial Intelligence and psychoanalysis set the context in which professional psychologists and the amateur psychologists we all are think about thinking. (p. 265)

REFERENCES

Benedikt, Michael (1992a). Introduction. In Michael Benedikt (Ed.) *Cyberspace First Steps*. Boston, MA: The MIT Press, pp. 1–26.

Benedikt, Michael (1992b). Cyberspace: Some Proposals. In Michael Benedikt (Ed.) *Cyberspace First Steps*. Boston, MA: The MIT Press, pp. 119–224.

de Certeau, Michel (1987). Practices of Space. In Marshall Blonsky (Ed.) *On Signs*. Maryland: The Johns Hopkins University Press, pp. 122–145.

Easlea, B. (1983). *Fathering the Unthinkable: Masculinity, Scientists and the Nuclear Arms Race*. London: Pluto Press.

Figueroa-Sarriera, Heidi (1991). *Metáforas de persona en textos sobre Inteligencia Artificial y Robótica*. Doctoral Dissertation. Dept. of Psychology, University of Puerto Rico, Río Piedras Campus.

Figueroa-Sarriera, Heidi (1996). Children of the Mind with disposable bodies. In Chris Hables Gray, Heidi J. Figueroa-Sarriera, Steven Mentor (Eds.) *The Cyborg Handbook*. New York: Routledge.

Figueroa-Sarriera, Heidi, and López, María M. (1993). Implicaciones del sujeto descentrado para las Ciencias Sociales o ¿dónde vives tu, finalmente?. *Revista Cayey*, Vol. XXIV, No. 72.

Foucault, Michel (1973). *The Order of Things*. New York: Random House.

Gergen, Kenneth (1992). *The Saturated Self. Dilemmas of Identity in Contemporary Life*. New York: Basic Books.

Gibson, William (1984). *Neuromancer*. New York: ACE Books.

Haraway, Donna J, (1992). A Cyborg Manifesto: Science, Technology, and Socialist-Feminism in the Late Twentieth Century. In Donna Haraway J. *Simians, Cyborgs, and Women. The Reinvention of Nature.* New York: Routledge. pp. 149–182.

Harvey, David (1990). *The Condition of Postmodernity.* Boston, MA: Basil Blackwell, Inc.

Heim Michael (1992). The Erotic Ontology of Cyberspace. In Michael Benedikt (Ed.) *Cyberspace First Steps.* Boston, MA: The MIT Press, pp. 59–80.

Lash S., and Urry, J. (1987) *The End of Organized Capitalism.* Cambridge: Polity.

Lefebre, Henri (1991). *The Production of Space.* Oxford: Basil Blackwell Ltd.

Malone, Robert (1978). *The Robot Book.* New York: Jove Publications, Inc.

Marvin, Caroline (1988). *When Old Technologies Were New. Thinking about Electric Communication in the Late Nineteenth Century.* Oxford: Oxford University Press.

Morningstar, Chip, and Farmer, Randall F. (1992). The Lessons of Lucasfilm's Habitat. In Michael Benedikt (Ed.) *Cyberspace First Steps.* Boston, MA: The MIT Press, pp. 273–302.

Novak, Marcos (1992). Liquid Architectures in Cyberspace. In Michael Benedikt (Ed.) *Cyberspace First Steps.* Boston, MA: The MIT Press, pp. 225–254.

Stone, Allucquere Rosanne (1992). Will the Real Body Please Stand Up?: Boundary Stories about Virtual Cultures. In Michael Benedikt (Ed.) *Cyberspace First Steps.* Boston, MA: The MIT Press, pp. 81–118.

Stenger, Nicole (1992). Mind Is a Leaking Rainbow. In Michael Benedikt (Ed.) *Cyberspace First Steps.* Boston, MA: The MIT Press, pp. 49–58.

Turkle, Sherry (1990). Artificial Intelligence and Psychoanalysis: A New Alliance. In Stephen N. Graubard (Ed.) *The Artificial Intelligence Debate. False Starts, Real Foundations.* Boston, MA: The MIT Press, pp. 241–268.

Virilio, Paul (1984). *L'horizon neqatif.* Paris: Editions Galileé.

Virilio, Paul (1988). *La machine de vision.* Paris: Editions Galileé.

Wexelblat, Alan (1992). Giving Meaning to Place: Semantic Space. In Michael Benedikt (Ed.) *Cyberspace First Steps.* Boston, MA: The MIT Press, pp. 255–272.

9
MANIFEST(O) TECHNOLOGIES: MARX, MARINETTI, HARAWAY
Steven Mentor

THE MANIFESTO AS MONSTER

By which writing technologies are technologies represented? And what are the politics of those writing technologies? These must be important questions for technohistorians; no one genre of representation determines the reception of technologies like electricity or automobiles, and below any essay on technology lie buried assumptions of what might constitute adequate and inadequate, normative and abnormal structures of representation. I've chosen to look at the representation of technology in manifestos because this genre appears to wear its politics on its sleeve, and because it conflates a particular view of technology with a highly self-conscious choice of stylistic and formal representation. This representation is itself a kind of writing technology built to shock as much as to persuade, to sell as much as to argue. For example, consider the difference between this paragraph and the following one.

All manifestos are cyborgs. That is, they fit Donna Haraway's use of this term in her own "A Manifesto for Cyborgs" - manifestos are hybrids, chimeras, boundary-confusing technologies. They combine and confuse popular genres and political discourse, borrow from critical theory and advertising, serve as would be control systems for the larger social technologies their authors hope to manufacture. Most include original ideas, but their aim is rather simulation, duplication, reproduction; they long to achieve the status of a rhetorical handgun passed out to masses of readers rather than that of a judge's scales. They are monsters of discourse, their de-monstrations reconstructing the audience (and their cultural landscape) in a strange and monstrous light; in Marinetti's famous phrase, they are made on the principles of violence and precision. They enact violence while pointing to the violence done by some Other/s; they use linguistic scalpels sharpened on the whetstone of newspaper headlines to disassemble and reassemble the body politic. Whether as homo faber and proletarian (Marx), Futurism's New Man (Marinetti), or cyborg (Haraway), the reader undergoes radical surgery, emerging with new prosthetics, often technological, but always discursive.

The preceding sentences give a taste of manifesto language, as well as some of the "body parts" of the manifesto as a literary genre. This chapter will explore ways in which a manifesto is itself a technology as well as a discourse about the politics of technology and instrumental reason. Marxism, Futurism, and feminism have all attempted to theorize the role of technology in the modern world, and in doing so have attempted to disassemble dominant stories about technology and reassemble them in utopian and material ways. Each attempts to remake political identity, to retell history as the history of new techniques of production, and to linguistically embody and enact this remaking and retelling. In each case, language is self consciously a technique; perhaps the manifesto is simply an extreme example of the ubiquity of myth and narrative in all attempts at technohistories, and part of the politics of any theory of technology.

THE COMMUNIST MANIFESTO: MELODRAMA OF TECHNOLOGY

A spectre is haunting the manifesto—the spectre of its double, literature. A manifesto is never simply a call to action, but is also a rhetorical construction of the proper scenes of action, the roles taken by diverse actors, the script of actions hoped for and believed in. Often the manifesto constructs such actions in a way quite different from what is accepted or commonly practiced; in that sense it must argue for its premises. But at the same time it attempts to frame these premises, not as doubted or new ideas to be analyzed, but as themselves obvious, evident, "manifest." This framing allows the language of the manifesto to soar in its denunciations and assertions, to transcend careful, hedged elaboration of political or artistic "programs" for the more powerful registers of rage and incitement.

The very origins of the word provide a glimpse at some of its internal tensions and contradictions. Manifest means readily perceived by the eyes or understanding, obvious, apparent, plain; its

Middle English antecedent *manifestus* is a variant of Latin *manufestus,* that is, struck with the hand. Most of the early manifestos are proffered by sovereigns and governments, agents with the material power to strike physically in order to make their meaning apparent and plain. And the root *manus* would provide a Foucauldian with a treasure trove of disciplinary and authoritarian terms: manage, manacle, manners, mandatory, mancipate (the power to sell slaves and other property), manipulate, command, demand, manuscript, and manufacture. (Partridge, 378–9).

But we have come to understand manifesto in an apparently opposite way: as emancipatory, a blow at some managing, and commanding authority, a slap at bourgeois and literary manners. Manifestos have come to signify the words of those outside the power to command: avant-garde artists, small, marginalized political groups and communities, individuals. They are metaphorical, rhetorical slaps of the hand by those who do not have the social or political power to proffer these slaps literally. As such, they are also attempts to manifest something that is not obvious; even the most "materialist" of manifestos thus shares something with the manifestation of spiritualists: the bringing to light of something that is immaterial, or immanent, in the consciousness or lived reality of humans in a specific society and historical period.

Why probe such elements of the manifesto? I certainly do not want to empty out the political contents or affective power of documents like the Communist Manifesto, or pretend that they are "merely" aesthetic constructions. In fact, this way of talking about value itself reproduces the problem I want to investigate. Just as literary productions have politics, have subversive or important effects on the symbolic economy of a society, so too political rhetorical productions have linguistic politics, and reflect important assumptions (about the nature of the political order, the roles and values of those opposed to it, right action) that have everything to do with the nature of subsequent material actions, or as the French say, "*manifestations.*" Further, manifestos not only make manifest certain kinds of actions and political organization; they obscure or evade internal contradictions or difficulties of the authors. I want to argue that these evasions or lacunae to play themselves out in very material ways, in subsequent actions and organizations engendered by the manifesto and its authors/signers. In the Communist Manifesto, "materialism" hides its own ghost, idealism and faith in the organic machine; the historical narrative of proletarian victory hides the attempts of nonproletarian intellectuals to shape and determine this victory and its means; the apparently bald and naked shape of the manifesto as clear argument and "realism" hides the equally powerful framing discourses of melodrama and catechism. And all these, I argue, play themselves out in the material nature of communist movements, states, attitudes toward technology and science, and in the kinds of political narratives and artistic movements they themselves are forced to exclude and even destroy. Every manifesto has its manifest, its bill of lading, which those unable to critically read it are doomed to pay and repeat.

When the Communist League called on Marx and Engels to draw up a manifesto in 1847, both wrote drafts. Engels' draft, titled "The Principles of Communism," is a catechism of twenty-five questions and answers, and the catechism is one of the genres rewritten here. But a catechism is a bald assertion of authority, one often backed up by the slap of a hand, as many who have attended Catholic schools can testify.[1] Engels's catechistic hand can most readily be seen in section two, "Proletarians and Communists," where the need to differentiate communism from other sects is most crucial. But most scholars agree that Marx is the main author of the present manifesto, and that the substance of the manifesto's narrative on history is his. The Marxian master narrative of history and technology is set against the "nursery tale" told about communism by its opponents; it is at the same time an argument for and demonstration of what seems clear, manifest, about history. Not only is this history told with simple, declarative sentences void of any hedging or qualification; the narrative itself is full of tropes of clarity, of the rending of veils and mystifications. Ironically, much of this work is done in the text by the bourgeoisie; in Marx's eyes, they perform the violent task, which he will continue, of demystification:

> [The bourgeoisie] has drowned the most heavenly ecstasies of religious fervor, of chivalric enthusiasm, of philistine sentimentalism, in the icy waters of egotistical calculation... The bourgeoisie has stripped of its halo every occupation hitherto honored... The bourgeoisie has torn away from the family its sentimental veil... (Tucker, 475–6)

Before, exploitation was "veiled" by "illusions" - now it is "shameless, direct, brutal." And this is in fact the rhetorical aim of the manifesto: to further this work by writing a "realistic" history. In case his readers miss this, Marx takes pains to connect intellectual production to material production and economy, and to point out that both are in the interests of the ruling class. Because of the rule of exploitation, "the social consciousness of past ages, despite all the multiplicity and variety is displays, moves within certain common forms . . . which cannot completely vanish except with the total disappearance of class antagonisms" (Tucker, 489). By implication, the present text attempts both a representation of such disappearance, and also a linguistic act that escapes these common forms that contain within them the traces of ruling class ideas. Hence the final assertions that Communists like the author "disdain to conceal" and "openly declare."

It is but a short step from this rhetoric to its stylistic predecessor and model: the "scientific" style adopted by the Royal Society as most appropriate for scientific inquiry and assertion. Marx was keen to claim for his socialism the title of scientific, and used this as a club with which to beat other forms of socialism as unscientific, romantic, utopian nursery tales. And yet the elements of the manifesto that go beyond mere catechism display certain common "unscientific" forms of its age, which shape its agenda and analysis. Coral Lansbury argues that the manifesto is based more on nineteenth century melodrama than on economic and historic discourses; I would add that melodrama is in fact a ghost that haunts many discussions of science and technology.

How is the manifesto a melodrama? The Communist Manifesto begins and ends with the signature: the Gothic ghosts that populate the works of August von Kotzebue, Francois Rene Pixecourt, and "Monk" Lewis. In its first English translation[2] the opening reads "A frightful hobgoblin stalks throughout Europe. We are haunted by a ghost, the ghost of Communism" (Lansbury, 6). Ghost, hobgoblin, "spectre" in the more familiar Samuel Moore translation: seen as malign by most, these spirits invoke a spiritual authority in nineteenth century melodrama: "The idea of a ghost as the moral conscience and protagonist comes directly from Gothic melodrama where the occult resolved its destiny through the mundane events of a historical present . . . the essence of the classic melodrama [includes] its benign ghost, its violent action, and the final social revolution, when the rightful heirs are restored to their proper place in society" (Lansbury, 6–7). An important element of melodrama is its manichean nature: all villains are aristocrats, all heroes and heroines are lowly born, noble peasants who are revealed as the true aristocrats. Many of these heroes literally cast off their chains in the end, as villains are often stymied at the last moment by the reappearance of the ghost; many heroines are the object of lecherous and rapacious squires and lords; most melodramas ended with sword play and the redistribution of spoils. For each of these genre-based elements, Lansbury cites passages from the manifesto: each element works on the emotions of the audience, to justify the social violence that destroys the demonic villain, to offer a world purged and restored to justice and order.

Lansbury's analysis is important for several reasons. First, it sets the manifesto more accurately within its time, and serves to deconstruct the status and class based antecedents often offered for it. Traditionally texts are made into unified monuments by citing the important and high status texts and authors on which it apparently draws (for Marx, obviously Hegel and Schelling); this analysis serves to counter this monumentalizing, to allow for multiple readings by citing the lower status, but arguably important antecedents that make a text popular or even conceivable in its day. The text is ineluctably hybrid. Beyond this, Lansbury offers another interpretation: the manifesto follows the melodramatic "logic of the excluded middle" because Marx and Engels attempt to avoid an obvious problem of being in the middle. That is, how is it that men from bourgeois backgrounds come to identify with the proletariat and speak for it?

> Marx and Engels showed no interest in understanding how and why intellectuals become radicalized . . . It was from the outset a problematic situation, for if social economic conditions inevitably determined historical change and human character, how was it possible for two members of the bourgeoisie to become heralds and spokesmen for the proletariat? . . . the process by which the bourgeois becomes a revolutionary intellectual and a standard bearer for the Vanguard Party relies more on faith than it does on factual analysis. (Lansbury, 3)

This evasion has had enormous effects on politics of the twentieth century. Lenin addressed it by theorizing a vanguard party of professional revolutionaries whose commitment to revolutionary violence masked their bourgeois origins; these intellectuals later became key players in purges of "bourgeois" intellectuals, journalists, labor leaders, as well as internal purges. Marx's guilt and resultant tale of purity is replayed in the Soviet Union, and later within the ranks of the New Left, where the "politics of guilt" allowed so-called revolutionaries from the Progressive Labor sect to take over SDS from its "bourgeois" student members.[3] Both Marx and Engels, in the manifesto, are concerned to discredit other forms of socialism, and especially the so called "utopian" socialism of Fourier, Owen, and others by labeling them unscientific, undeveloped, fantastic, indistinct. Marx denounces their "castles in the air" based on "their fantastical and superstitious belief in the miraculous effects of their social science" (Tucker, 499). Other sects are similarly denounced. This "excluded middle" of activism and socialist activity repeats much of the same rhetoric of religious sectarianism, and a similar appeal to purity. If melodrama is "a mode of compulsive seriousness seeking to restore a fragmented society to a new and harmonious whole" (Lansbury, 4) then Marx repeats on the revolutionary stage the same romanticism and naivete he denounces in bourgeois society. Melodrama is the signature of a powerful desire arising from material conditions, but few would argue that it provides a realistic or scientific model of action for a millennial proletarian revolution. And melodrama's obligatory violence is transferred to the discourse of actual political change, so that violence becomes a mark of purity, and its lack a sign of "bourgeois" decadence or armchair socialism. Hence notions of revolutionary resistance that don't include revolutionary violence or terror as integral elements are banished to the "excluded middle."

One of the ghosts that haunts this rhetorical machine is technology. Technology in the hands of the bourgeoisie is violent and vengeful: it tears, drowns, and then establishes a new naked form of exploitation. Yet for Marx it is absolutely necessary that technology play this role: the violence of the Industrial Revolution, the internationalization of capital, "the constant revolutionizing of production," and "uninterrupted disturbance of all social conditions," are crucial for a materialist epiphany that moves beyond religious and feudal myths. This necessity is placed next to phrases on the deskilling of craftsmen, enslavement to the machine, the devolution of humans to commodities, "appendage[s] of the machine." We are all familiar with the final lines of proletarians having nothing to lose but their chains; however, most of the energy in Marx's prose lies with the technological forces of production, which burst chains (in the Tucker, fetters) repeatedly in section one.

By making the bourgeoisie a revolutionary class, by lending forces of production monstrous and unstoppable agency, and by constructing a melodramatic analysis of technology, Marx paves the way for the unquestioned Fordism of Lenin and Gramsci, and the precedence of industrial power and statist control over the lived relations of workers and their tools/machines. The Communist State and its vanguard leaders will unfetter first and foremost technological and industrial production; if it is true that bourgeois culture for the worker is "a mere training to act as a machine" (Tucker, 487), how will State socialist culture, with its reverence for the exact same industrial methods, be any different? All that is solid—the worker's felt alienation and anger toward his commodification and mechanization—is melted by the middle class Marxist rhetoric into air.

Yet we should also acknowledge the enabling effects of Marx's melodrama of communism. It certainly appealed to the audiences of his time, and arguably to many audiences in the twentieth century. It enacts on the rhetorical level the notion of drama, of conflicts based on recognizable present day historical roles, events, and genres. And it serves as an affective gateway to a Marxian narrative that includes truly radical and powerful revisions of history, economics, technology, classes, and the state. Whether consciously or not, it blurs the boundaries between history and drama, economics and the conventions of fiction, the familiar roles in art and the unfamiliar roles of Marxian political landscapes. It also serves as the Father against which later authors of manifesti, including Marinetti, Breton, and Haraway, would both rebel and measure themselves.

DRIVING IN THE DARK: MARINETTI'S DEUS EX MACHINA

Marx's manifesto generated some strange and rebellious offspring. In her book *The Futurist Moment* Marjorie Perloff produces a narrative about manifestos as a literary genre, beginning with the Communist Manifesto, reaching a kind of apex in the Futurist works of Marinetti, Boccioni, Balla et al., and continuing into Dada and Surrealist manifestos of the period. Perloff finds in the Communist Manifesto's "curiously mixed rhetoric" (which she sees as a prose poem) a model of what the manifestos of the twentieth century will do: graft poetic onto political discourse. (Perloff, 82). Other early twentieth century artists and thinkers such as Saint-George de Bouhelier, Jules Romain, and Ernst Ludwig Kirchner, anticipate themes of Futurist and later manifestos: the attack on symbolism, the urge toward energy and violence, urban mass art, and the ever-present need to create a new art and literature. But these manifestos remain formally similar: they begin with generalizations about art, and generally follow a nineteenth century model of oratory and persuasions, marshalling arguments, balancing emotional appeals with reasoned and extended discourse. And few deal specifically with new technologies.

By contrast, Marinetti's 1909 manifesto (Fondation et Manifeste du Futurisme) begins with a narrative that sings the body electric and makes new technologies the key to artistic and political rejuvenation. Marinetti and his friends have stayed up all night "arguing to the last confines of logic" and scribbling. Thus the logics of previous manifestos are exhausted within the first paragraph; instead, like Marx's unfettered technologies, "the prisoned radiance of electric hearts" is freed by the call of mechanical sirens: great ships, locomotives, huge double decker trams, and most of all, Marinetti's car (machina). Just as Marx uses melodrama, Marinetti marries a late symbolist aesthetic to technology: his car is a beast, a dog biting its tail, a prodigy, a shark; and presaging so many technophilic American movies, this 1909 piece begins with a car crash and ends with the wholesale destruction of the venerable city with its dead museums and academies. The car crashes when Marinetti swerves to avoid two bicyclists "wobbling like two equally convincing but nevertheless contradictory arguments" (Flint, 40).

Joy of the machine versus logic and paralysis; the feeling of having avoided death. The reader is in a rhetorical machine that uses the resources of the symbolist and prose poem to join human virtues (courage, audacity, energy) to technology: "A racing car . . . is more beautiful than the Victory of Samothrace." Where Marx saw the worker enslaved by the machine, Marinetti puts him at the wheel of a car; the reader is accelerated through violent and extreme positions, not pausing to wonder whether or how artistic revolution goes with glorifying war, or how "art in fact can be nothing but violence, cruelty, and injustice." Even the most programmatic elements (numbered theses, for example) are swept aside in a verbal vortex that performs its message. Engels's catechism is jettisoned and parodied; not argument and principle, but "de la violence et de la precions"[4] will be the principle of Futurist manifesti.

Of course, this is still a time when most workers are chained to their machines, figuratively if not literally (as for example many women seamstresses were); few could afford the new wheeled machines of the millionaire Marinetti. Yet in some important ways the Futurist manifesto is related to car advertisements and the language of publicity. Marinetti figuratively says to his car, "I love what you do for me," and like so much ad copy, he uses a brief and dramatic visual story (a car crash) to set the pace and tone of the pitch. Note that the pitch mingles the new consumer item (cars) with images of vital industrial society in general (railroad stations, factories, shipyards): it sings "of great crowds excited by work, by pleasure, and by riot" as if the work of industrial society was homologous with the symbolic energy of "deep chested locomotives whose wheels paw the tracks."

In fact Marinetti was a tireless promoter who travelled by rail from agitation to agitation; he posed his followers for publicity stills, managed to get his manifesti printed on front pages of French and Italian journals, and perfected the public scandal: attacking a Venetian orchestra, setting up mechanical altars to the Fatherland in squares, insulting and then cajoling crowds until fights broke out. Like the first Parisian performance of Rite of Spring, which ended in a shouting match and riot, Marinetti found a way to wed avant garde theories of art and writing with new competencies at marketing, publicizing, and distributing his message. The fact that technology and technophilia is at the heart of this message is not surprising, since it is the technology of publicity and movement that allows his small group to gain such widespread notoriety and power. And this technophilia is intimately wedded in Futurism to glorification of war (now itself dominated by technology) and masculinist nationalism: "We will glorify war—the world's only hygiene—militarism, patriotism . . . scorn for women" (Flint, 42).

Each element of this publicity will be used by fascist and Nazi movements to mobilize the masses and gain state power: not just attacks on the bourgeoisie and the status quo, but the invocation of a potent future based on the promises of technology and manifested through the language and techniques of modern publicity. The technology of advertising and the advertising of technologies combine to form a powerful and exciting narrative of progress. Like Marx's unfettered technology, the engines and aeroplanes of Futurism stand for an almost magical force that counteracts the routinization of modern life and labor. But where Marx saw this force as inevitably international in scope, Marinetti and Futurism tend to link technology to the body of the nation-state, made strong by war's hygiene, alive by state electrification grids, pleasurable by the speed of highways, sublime by the power of gigantic industrial dynamos. The aesthetisation of technology hides it political uses and the continuity of deskilled mechanized labor under fascism no less than communism.

GENDER TROUBLE: KEEPING THE MAN IN MANIFESTO

Marinetti's manifesti raise questions about the gender of political rhetorical machines. Cinzia Blum has analyzed the rhetorical strategies of Marinetti's Futurist manifesti in the process asking: to what extent does the apparent revolution in style and genre signal a parallel revolution in political action? Does the manifesto's subversion of traditional codes, genre boundaries, and expressive registers mean also "the undermining of hierarchical, centralizing, ordering systems predicated upon a unitary, authoritative speaking and thinking subject"? (Blum, 197).

By looking closely at the language and binary oppositions of Marinetti (in "Fondazione e Manifesto del Futurismo") and his followers, Blum discovers a response to the anxieties of modern(ist) fragmented identities and social codes that should not surprise us: anxiety and self-doubt are erased, in Futurism, by the demonizing of feminized Others and a recuperation of phallic mastery via fantasies of omnipotence and sexual aggression.[5] The construction of this "fiction of power" is a compensation for the lack of such power in the modern world. She uses Kristeva's notion of the abject to link strong separation of the sexes with fear of that which traverses the boundary of the self; while Futurism appears to theorize the destruction of the unitary self of previous literary and political constructions, it recuperates this potent self as the "multiplied man":

> In fact, the scattering ("sparpagliamento") of the self in the universe (brought about by the fast pace of modern life) is presented as a means to a more powerful unity freed from the limits of human nature . . . the Futurist subject disperses himself to penetrate the molecular life of matter, and with aeropoesia, rises as a super "I" propelled by mechanical wings to control immense spaces in the totalizing . . . perspective allowed by the airplane. (Blum, 204)

In the process of recuperating the virile and potent male subject, various and sundry "others" must be overcome, indeed penetrated and destroyed. The site of violent action is the manifesto, but also women's bodies and the things they stand for: impotence, disease, fragmentation and powerlessness, chaos and the undefined, love and the limits of human/nature, the decadent, the organic; parliamentarism, pacifism, academic culture, psychological writing. In the face of so much experimentation by Futurists, one barrier remains policed: gender in language. One Futurist, Francesco Canguillo, actually argues that sexual perversion may result from linguistic perversion, and that reducing ambiguous grammatical gender will simultaneously fix meaning and deviance: "Although other linguistic rules can and must be subverted in the name of artistic freedom, or rather, of the artist's power, grammatical gender is the object of reactionary, homophobic concerns, of an effort to restore the oldest conception of language—that of the intrinsic relation between signifier and signified" (Blum, 199). Blum argues that the Futurist's emphasis on masculine culture managed its undercurrent of homoerotic desire by displacing this homoeroticism onto the machine, and, I would add, onto the literary product as a machine and a site of mastery over feminized others.[6] Ultimately, "while the manifesto's hybrid nature instantiates the disruption of codes in modern chaotic, fragmentary reality, the rhetoric and thematics of gender strive to establish more rigid gender codes which provide for the integrity of the subject and for an unwavering code of authority and subordination" (Blum, 200).

The Futurist movement generated many opposing manifesti: Mina Loy's feminist manifesto appropriated Futurism for feminism and attacked Marinetti's sexism using his own terms; Dadaists like Tristan Tzara and Surrealists like Andre Breton used the manifesto form to attack his militarist and nationalist views of technology and his use of avant garde technique to defend reactionary and fascist modes of social organization. Yet it remained for feminist theorist extraordinaire Donna Haraway to bring gender, technology, and politics together in the cyborg mother of all manifestos.

A CYBORG FOR MANIFESTOS: READING DONNA HARAWAY

As I hope I have shown, the manifesto is already a cyborg; Donna Haraway's 1985 "A Manifesto for Cyborgs" can be read as a redundancy, a manifesto for manifestos, a guide for writing politically charged histories of technology and feminism.[7] If the Communist Manifesto had remained Engels's catechistic discriminations and a taxonomy of nineteenth century socialisms, we would not be reading it today; and if Haraway's 1985 article had limited itself to a critique of totalizing feminist and socialist narratives, or to a weave of feminist theorists and postmodern economics, it might never have left the predictable orbit of *Socialist Review* and its readership. Instead, Haraway rewrites Marx via avant garde manifesto strategies of Marinetti, Breton, and Guy Debord; like Marinetti, she uses violence as well as precision to achieve a powerful analysis of technology and politics in the late twentieth century.

We have seen the rhetorical violence Marx deploys when he invokes the rending and tearing of veils accomplished by the bourgeoisie and their industrial technologies; in his manifesto, both the vital force of technological progress and the coming solidarity of the proletarians burst fetters, haunt a terrified ruling class; and a dominant metaphor is war, the war of class against class. This way of

Manifest(o) Technologies

seeing social relations and technology is not hedged; though other paradigms are possible, Marx performs the notion of war by simultaneously claiming to describe and declare war. Often, commentators have noticed the contradiction between a professed state of war and a strategy that depends on building workers' parties within the political structures of bourgeois society. And we have seen the dilemma of maintaining that industrial methods that enslave workers will ultimately free them.

A similar dilemma—Marinetti's two bicyclists threatening logical paralysis—inhabits Haraway's piece. Beyond the Marinetti-like witty insults (creationism for example is described as "child abuse") Haraway describes a "border war" within "racist male dominated capitalism" and its sciences: a war over the borders of organism and machine, whose stakes, like those of Marx, are production and imagination, and unlike Marx, involve reproduction. Haraway both discovers and enacts this violent border war; like Marinetti crashing the reader/passenger into the industrial muck, like Marx disassembling the image of the organic society and the craft worker, she forcibly situates us: "we are cyborgs."[8] On one level this simply refers to the nature of late capitalism: she argues for a fundamental change, "an emerging world order ... a movement from an organic, industrial society to a polymorphous, information system." In this new world dis/order, the makers of material cyborgs—the military, industry, medicine—all reduce the "human" to parts within a larger cybernetic system that includes machines. To give her readers the feeling for this reality, she ironically deploys the language of engineers and systems:

> In relation to biotic components, one must think not in terms of essential properties but in terms of strategies of design, boundary constraints, rates of flow, systems logics ... Any objects or persons can be reasonably thought of in terms of disassembly and reassembly; no "natural" architectures constrain system design ... Human beings, like any other component or subsystem, must be localized in a system architecture whose basic modes of operation are probabilistic, statistical. No objects, spaces, or bodies are sacred in themselves; any component can be interfaced with any other is the proper standard, the proper code, can be constructed for processing signals in a common language. (Haraway, 594)

This language performs its own violence; try as we might, it is difficult to think of ourselves as bounded organic individuals within such a field of discourse. And this is the discourse that governs the political and technical world Haraway wants us to inhabit.

Haraway deploys another type of violence: the violence of precision. Her opening section relentlessly piles on multiple definitions of the cyborg, refusing to change register or descend to illustration, development, explanation. Like Marinetti, she knows the power of speed and substitution. If "we are all cyborgs" then these fast-shifting definitions all somehow apply to us, no matter now diverse. It is exhilarating to imagine that a technological shift, one which batters down socially constructed boundaries of organic humans and mechanical machines, could have such futuristic and utopian effects: "we" are thus in a postgender world, beyond false unities and false origin stories, heterosexual and patriarchal expectations, with a natural feel for united front politics. And besides these laudable feminist qualities, we also are monstrous, capable of bestialities, always multiple and incomplete. These latter qualities are also effected by her language; it disassembles us as organic and reassembles us as a proliferation of qualities which do not easily fit any whole or synthesis or even politics. We cyborgs are torn apart as by maenads, spread across a discursive field, mingled with various technologies and discourses (C^3I, late capitalism/economics, feminism, socialism, poststructuralism, literary theory) and thus capable of any number of assemblages.

If Haraway rhetorically reembodies us as cyborgs, she also makes our cyborg selves visible. This is in fact a trope of manifesti: metaphors of sight, of disclosure, of making the invisible visible, run through Marx, Marinetti, and the others. "The ubiquity and invisibility of cyborgs is precisely why these sunshine-belt machines are so deadly. They are as hard to see politically as materially" (Haraway, 584). In Haraway's postmodern melodrama, cyborgs haunt not only Europe and its humanist legacy, but also left and other oppositional groups who find it hard to confront borderless transnational corporations, science always already implicated in military research, political and technological maps based on systems theory and invisible flows of data over networks whose bodies are at once human and mechanical and electronic. Haraway's point in violently resituating us: the tendency of progressives to confront the "domination of technics" with "an imagined organic body to integrate our resistance" misses the increasingly hybrid, cyborgian nature of our lived bodies and societies. Marx critiqued socialisms which failed to "see" the ubiquity and dynamism of industrial techniques; Haraway critiques oppositional groups (Marxist feminism and radical feminism) which fail to take into account the "informatics of domination" based on cybernetic and communication systems, neo-imperialism and neo-colonialism.

THE HYBRID AS HYBRID: RHETORICAL FEEDBACK LOOPS OF REFLEXIVE NONTOTALITY

Both Marx and Marinetti construct technology as a "vital machine," hybrids of organic and machinic ways of looking at the world. As David Channell writes, many post-Romantic nineteenth century writers revived an organicist mode of looking at the world, including technology:

> ... an opposing organic world view ... used the symbol of an organism, such as the body or a plant, to understand the world ... For the organicist the organization of parts into a whole result in qualitatively new phenomena such as a vital spirit principle or force ... In such a world view there is also no conflict between machines and organic processes since both will be thought to arise from some vital organization. (Channell, 9).

Hegel's figure of the bud that flowers and Marx's appropriation of this type of image for the force of technology, are typical. By constructing railroads, telegraph systems, and other communications technologies, along with industrial modes of organization that allow new communication of misery and solidarity between workers, the bourgeoisie unwittingly build an "organic machine" on the scale of society, which will literally manufacture the proletarian or Futurist class. And Marx's manifesto is also such an organic machine: the sum of its analytic parts are greater than bourgeois society, greater than any demonstration or proof. The melodramatic ghost in the technological machine breaks all mechanistic fetters, all attempts by bourgeois society to contain it.

Haraway replaces the notion of holism and organic machine with the figure of the cyborg; in this she is joined by theorists such as Channell and Bruce Mazlish. Mazlish sees the human/machine boundary breaking down and providing a fourth great discontinuity to our conception of the human[9]; Channell suggests that artificial intelligence, genetic and biomedical engineering reflect a watershed merging of mechanism and organicism into a bionic world view:

> Unlike the reductive approach of the mechanical view or the holistic approach of the organic view, the bionic world view is consciously dualistic in its understanding of the world ... [which] emphasizes the role of interactive processes or dualistic systems in understanding the world. (Channell, 10)

If the organic machine circulates through Marx and Marinetti, it also circulates as a narrative of holism and necessity. Thus Marinetti can call war "hygiene." Thus Marx can with utter assurance give us an etiology of socialism that rejects amputated or excessive bodies of knowledge as literally diseased, while retaining health and bodily coherence for his own ideology. By contrast, Haraway deploys the cyborg to do more than point at new intimacies of technology and human; she attacks the discursive claims to holism and to vitality, to totality and total explanatory and motivating power, of organicist narratives. She does this partly by using familiar poststructuralist arguments, but also by demanding that we see her manifesto as both fictional and "real," both constructed and in some important sense vital, alive.

Calling attention to the fictional and assembled nature of her production, framing her manifesto with notions of myth and story and fiction, Haraway theorizes technology is similar fashion, not as organically developing but as assemblable and so disassemblable and reassemblable. Humans are part of and parts of social technologies; to the extent that the cyborg figure makes visible the blurred boundary between biotic and mechanic, between individual humans and technical systems, it allows humans to tell different, multiple stories about technology. And it gives those stories potential feedback loops and prosthetic rhetorical limbs: we might replace Haraway's discussion of science fiction with newer or different texts, or add an entire section on bioengineering and gender.

One important feedback loop in this manifesto concerns the figure of the cyborg and the limbs of analysis. If the organic machine centers Marx's text and gives it coherence, the cyborg both centers and decenters Haraway's text. In true manifesto form, she enacts "the" cyborg, in all its utopian, violent, monstrous possibility; yet the excessive list of qualities could easily continue. The cyborg is a defining figure, one which dramatizes our imbrication within technical systems and allows us to rethink dualisms about humans and technology; but it is itself inherently capable of many transformations. Marx proclaims his theory of technology inevitable and scientific; Marinetti ends his masculinist manifesto "Erect on the summit of the world" and sees from airplanes, the God perspective. Haraway rejects this God's eye view and forces us as readers to negotiate the blurred boundary between science fiction and fact, myth, and analysis. This rhetorical cyborg for example involves the cybernetic discourse systems of feminism and materialism; we could as easily build a "central" cyborg out of military or medical discourses. The latter might similarly confuse gender constructions, but with arguably different effects.

Manifest(o) Technologies 211

This feedback has discomfited more than a few readers. How can Haraway describe what "the" cyborg means with such confidence, and yet make statements like "who cyborgs will be is a radical question"? Or how can a reader understand "the" cyborg if asked to take two perspectives, one "the final imposition of a grid of control on the planet . . . the final appropriation of women's bodies in a masculinist orgy of war" and the other a lived experience of partial identities and kinship with animals and machines? The key is in the notion of performatives and textual machines. Haraway indicates that the figure of cyborg works for seeing many elements of twentieth century technology and politics; if indeed a cyborg is product of variable systems, then Haraway persuades us to inhabit more than a couple of cyborg bodies during the course of the essay. None is "necessary"; none makes everything whole or complete; the multiple shifts make a mockery of all consuming taxonomies and inevitable trajectories of technical development.

Thus Haraway's cyborg manifesto contains a cyborg writing that joins the reader to different prosthetic rhetorical machinery. She imagines the aeropoetic pleasures of a Marinetti joined to the social responsibility of a Marx, while inviting the reader to see technologies and rhetorics as discursive, narrated, rewritable. Her manifesto implicitly critiques all manifesti that attempt to hide their discursive and mythic status, while arguing for the engaged and political nature of all representations of technology.

CONCLUSION

> The cyborg is the figure born of the interface of automaton and autonomy.
> —Donna Haraway, *Primate Visions* (139).

If writing is a technology, then writing about technology, writing technohistories, demands a doubled vision. Cyborgs and other technologies as discursive, chimerical, mythical objects circulate in the least likely places: government policy statements, military research and development reports, medical journals. They carry with them narratives, perspectives, genres, that belie the staid generic prose of their textual bodies. Even a cursory look at the history of attempts to represent technology and its social implications must surely reveal that all such attempts are always already mythical, narrated, fictional; bringing these ghostly figures to light in current technohistories must be a prime goal. This is not to enter the slippery slope of relativism, in which all texts are equally false or suspect; rather, it is to suggest that nuclear power plants, waste management systems, and medical cyborgs all escape any one genre of representation, comedy or tragedy, romance or farce. We must look at the institutional and political interests embedded in such generic representations, as well as our own framing stories and technologies of representation. Initially, this may be giddy, unfamiliar business, rather like the figure/ground reversals of avant garde collage, the fragmentations of cubism.

Artifacts indeed have politics; technologies can be agents. We want to think of organic humans making autonomous decisions about humane uses of technologies, but instead we must learn to think of humans and machines linked in multiple, often invisible, networks and systems of power. The discursive systems used to represent such systems are part of the system, but they are not the whole system; human bodies, wills, and stories do not consciously rule these systems (autonomy), but neither are they absent (automaton). Every technohistory constructs what it pretends to discover, performs what it pretends to demonstrate; Haraway's great gift to both political and technological history is to acknowledge that this is always so. Humans are radically constrained and constructed by the technological systems developed up to now; as John Christie points out in his "A Tragedy for Cyborgs,"[10] the future is in certain ways already written. Yet as I write this, technologies like the Internet, bioengineering, genetic research, and expert systems are undermining basic tenets of political bodies/technologies like nation-states and their governmental apparatuses. These organic machines and their legitimizing stores will be transformed in ways impossible to imagine now; more manifesti wait to be written.

NOTES

1. This is ritualized in Catholic ceremonies of confirmation, during which in my case each kneeling boy was lightly slapped by the bishop.

2. The first English translator is Helen Macfarlane, whom Lansbury describes as an early militant woman journalist and novelist. Her style, more than that of the later Moore translation, seems to preserve the melodramatic flavor, especially in this opening and in sections concerning women.

3. The latter is well documented in Sale, *SDS*. An important manifesto countering this excluded middle is "The Politics of Guilt" by Greg Calvert, one of the leaders of SDS cited in Sale.

4. Marinetti's advice to painter Henry Maassen, cited in Perloff, p. 81.

5. For an extreme example of this, read the later stories of D. H. Lawrence, especially "The Women Who Rode Away" and *The Plumed Serpent*. Lawrence was heavily influenced by Marinetti.

6. Klaus Theweleit comes to similar conclusions in his study of postwar German male attitudes toward war and machines in his *Male Fantasies*. In his analysis, men of the World War I Freikorps inhabit "the conservative utopia of the mechanized body" (2:162). Marinetti's genius was in spotting the consumerist mechanized body in advance of the fascistic, warrior version.

7. My citations from this article refer to the reprint in Hansen and Philipson, eds., *Women, Class and the Feminist Imagination*. This is an exact reprint of the original article from *Socialist Review*, No. 80, March-April 1985.

8. A cyborg is a cybernetic organism. Originally coined to describe possible bioengineering of astronauts for living in space (Clynes and Kline, 1960) the word now generally applies to more or less intimate connections between machine systems and organic bodies which include feedback mechanisms and homeostatic parameters of operation.

9. Freud refers to three great shocks to the human ego—Copernicus, Darwin, and himself—in his *General Introduction to Psychoanalysis*. Cf. Mazlish, pp. 3–5.

10. Christie links Haraway and William Gibson as proponents of a technological sublime, and argues that Haraway underestimates the unpleasant effects of virtual reality, late capitalism, and cyberspace on a disappearing organic body and on debtor Third World nations. I believe Christie's points are well taken, and yet he seems to generically misread the manifesto's stated goals of irony and blasphemy. Haraway is well aware of the tragic reading of technology and the classic humanism such tragedy presupposes; she just doesn't believe such a reading gives us the resources to theorize and redirect potent new technologies.

REFERENCES

Blum, Cinzia. "Rhetorical Strategies and Gender in Marinetti's Futurist Manifesto." *Italica*, vol. 67(2), Summer 1990, pp. 196–211.

Channell, David. *The Vital Machine: A Study of Technology and Organic Life*. Oxford: Oxford University Press, 1991.

Christie, John. "A Tragedy for Cyborgs." *Configurations,* vol. 1(1), Winter 1993, pp. 171–196.

Clynes, Manfred and Nathan Kline. "Cyborgs and Space." *Astronautics,* September 1960.

Flint, R.W., ed. *Marinetti: Selected Writings.* Translated, R.W. Flint and A. Coppotelli. New York: Farrar, Strauss and Giroux, 1971.

Haraway, Donna. "A Manifesto for Cyborgs: Science, Technology and Socialist Feminism in the Last Quarter." In *Women, Class and the Feminist Imagination.* Edited by Karen Hansen and Ilene Philipson. Philadelphia: Temple University Press, 1990.

Haraway, Donna. *Primate Visions: Gender, Race and Nature in the World of Modern Science.* New York: Routledge, 1989.

Lansbury, Coral. "Melodrama, Pantomime, and the Communist Manifesto." In *Browning Institute Studies: An Annual of Victorian Literacy and Cultural History,* Volume 14, pp. 1–10. New York: The Browning Institute, 1986.

Mazlish, Bruce. *The Fourth Discontinuity: The Co-Evolution of Humans and Machines.* New Haven: Yale University Press, 1993.

Partridge, Eric. *Origins: A Short Etymological Dictionary of Modern English.* New York: Greenwich House Press, 1983.

Perloff, Marjorie. *The Futurist Moment, Avant-Garde, Avant-Guerre, and the Language of Rupture.* Chicago: University of Chicago Press, 1986.

Sale, Kirkpatrick. *SDS.* New York: Random House, 1974.

Theweleit, Klaus. *Male Fantasies.* Translated S. Conway, E. Carter, C. Turner. 2 vols. Minneapolis: University of Minnesota Press, 1977–8.

Tucker, Robert C. *The Marx-Engels Reader,* 2nd Edition. New York: W.W. Norton, 1978.

10
NASA RETROSPECT AND PROSPECT: SPACE POLICY IN THE 1950s AND THE 1990s

Roger D. Launius

INTRODUCTION

With the end of the Cold War between the United States and the Soviet Union, advocates of an aggressive space exploration program face a problem similar to what they had encountered more than 40 years ago before the Sputnik crisis. Then, as now, space enthusiasts were motivated by an expansive view of human voyages of discovery, the exploration and settlement of the Moon and other planets of the Solar System, and eventual interstellar travel. The sense that we have to get off this planet, if the human race is to survive indefinitely, is a compelling aspect of this dream of space flight.[1] In their view the human component of space flight has been the central one, with robotic probes and applications satellites a useful but decidedly less important aspect of the space exploration agenda. But adventure and discovery, as well as the long-range goals of exploration and colonization, have never been rationales with enough attraction to justify for political leaders the ventures space advocates have wished to pursue.

This chapter will explore the role of analogy in space policy in the 1950s and the 1990s. Analogy is a useful tool in efforts to learn from history, and this seems to be an appropriate one to examine for people involved in public policy formulation related to the space program and for those interested in the history of the space program. A good many analogies, of course, are not appropriate, as in George Bush's comparison of Saddam Hussein's invasion of Kuwait to Hitler's incursions into the Rhineland, Czechoslovakia, and Austria before World War II. Hussein was a far cry from Hitler and Iraq was never comparable to Germany, and vice versa.[2]

While many analogies are useful analytical tools, they must be invoked with full awareness of their appropriateness and limitations. To determine their appropriateness and limitations, those using analogy as an analytical tool must define the immediate situation (the known) and the decision-makers' concerns (problems) with it, from which to draw objectives. When these are considered along with an explicit analysis of what is known, what is unclear, and what is presumed in the two situations being compared, the process can yield useful information. As public historians have been trying to convince policy-makers in Washington for many years, arguments based on hard "facts" about whatever issue is under consideration are convincing, but those same arguments based on historical data and the proper use of analogy can be overwhelming.[3]

The discussion of the space policy debates of the 1950s and 1990s seems an appropriate place to use analogy. In essence, this discussion focuses on three major points of comparison. The first is the nation's sense of adventure and discovery. The second is the popular conceptions of space travel and the possibilities it holds for Americans. The third is the role of international relations, foreign policy, and national security issues in setting the agenda for the space program. Although issues of Federal budgets and economic concerns, practical political interests, technological capability and vision, and a host of more obscure matters could also be discussed in detail, they have been analyzed at length elsewhere and need not be dealt with here.[4]

THE SPACE POLICY DEBATE IN THE 1950s
The Role of Adventure and Discovery

There seems to be little doubt but that adventure and discovery, the promise of exploration and colonization, were the motivating forces behind the small cadre of early space program advocates in the United States prior to the 1950s. Most advocates of aggressive space exploration efforts invoked an extension of the popular notion of the American frontier with its then attendant positive images of territorial discovery, scientific discovery, exploration, colonization, and use.[5] From Captain Kirk's soliloquy—"Space, the final frontier"—at the beginning of each *Star Trek* episode to Kennedy's speech about setting sail on "this new ocean" of space, the frontier allusion was a powerful component of early efforts to promote the space program. Indeed, the image of the American frontier has been an especially evocative and somewhat romantic, as well as popular, argument to support and aggressive exploration of space. It plays to the popular conception of "westering" and the settlement of the American continent by Europeans from the East that was a powerful metaphor of national identity until the 1970s.

The space promoters of the 1950s and 1960s intuited that this set of symbols provided a vigorous explanation and justification of their efforts. The metaphor was probably appropriate for what they wanted to accomplish. It conjured up an image of self-reliant Americans moving westward in sweeping waves of discovery, exploration, conquest, and settlement of an untamed wilderness. In the process of movement, the Europeans who settled North America became in their own eyes a unique people from all the others of the Earth imbued with virtue and justness. The frontier ideal has always carried with it the ideals of optimism, democracy, and right relationships. It has been almost utopian in its expression, and it should come as no surprise that those people seeking to create perfect societies in the seventeenth, eighteenth, and nineteenth centuries—the Puritans, the Mormons, the Shakers, the Moravians, the Fourians, the Icarians, the followers of Horace Greeley—often went to the frontier to carry out their end.

It also summoned in the popular mind a wide range of vivid and memorable tales of heroism, each a morally justified step of progress toward the modern democratic state. While the frontier ideal reduced the complexity of events to a relatively static morality play, avoided matters that challenged or contradicted the myth, viewed Americans moving westward as inherently good and their opponents as evil, and ignored the cultural context of westward migration, it served a critical unifying purpose for the nation. Those who were persuaded by this metaphor, and most white Americans in 1960 did not challenge it, recognized that the space supporters were summoning them not to recall past glories but to undertake—or at least to acquiesce in—a heroic engagement with the forces of social, political, and economic injustice.[6]

The Role of Popular Conceptions of Space Travel

If the frontier metaphor of space exploration conjured up romantic images of an American nation progressing to something for the greater good, the space advocates of the Eisenhower era also sought to convince the public that space exploration was an immediate possibility. It was seen in science fiction books and film, but more importantly, it was fostered by serious and respected scientists, engineers, and politicians. Deliberate efforts on the part of space boosters during the late 1940s and early 1950s helped to reshape the popular culture of space and to influence governmental policy. In particular, and this is the point I want to emphasize, these advocates worked hard to overcome the level of disbelief that had been generated by two decades of "Buck Rogers" type fantasies and to convince the American public that space travel might actually, for the first time in human history, be possible.[7]

The decade following World War II brought a sea change in perceptions, as most Americans went from skepticism about the probabilities of space flight to an acceptance of it as a near–term reality. This can be seen in the public opinion polls of the era. For instance, in December 1949 Gallup pollsters found that only 15 percent of Americans believed humans would reach the Moon within fifty years, while 15 percent had no opinion and a whopping 70 percent believed that it would not happen within that time. By 1957, 41 percent believed firmly that it would not take longer than twenty five years for humanity to reach the Moon, while only 25 percent believed that it would. An important shift in perceptions had taken place during that era, and it was largely the result of a public relations campaign based on the real possibility of spaceflight coupled with the well-known advances in rocket technology.[8]

There were many ways in which the U.S. public became aware that flight into space was a possibility, ranging from science fiction literature and film that were more closely tied to reality than ever before[9] to speculations by science fiction writers about possibilities already real[10] to serious discussions of the subject in respected popular magazines. Among the most important serious efforts was that of the handsome German emigre, Wernher von Braun, working for the Army at Huntsville, Alabama. Von Braun, in addition to being a superbly effective technological entrepreneur, managed to seize the powerful print and electronic communication media that the science fiction writers and film makers had been using in the early 1950s and no one was a more effective promoter of space flight to the public.[11]

In 1952 von Braun burst on the broad public stage with a series of articles in *Collier's* magazine about the possibilities of space flight. The first issue of *Collier's* devoted to space appeared on March 22, 1952. In it readers were asked "What Are We Waiting For?" and were urged to support an aggressive space program. An editorial suggested that space flight was possible, not just science fiction, and that it was inevitable that humanity would venture outward. Von Braun led off the *Collier's* issue with an impressionistic article describing the overall features of an aggressive space flight program. He advocated the orbiting of an artificial satellite to learn more about space flight followed by the first orbital flights by humans, development of a reusable spacecraft for travel to and from Earth orbit, building a permanently inhabited space station, and finally human exploration of the Moon and planets by spacecraft launched from the space station. Willy Ley and several other writers then followed with elaborations on various aspects of space flight ranging from technological viability to space law to biomedicine.[12] The series concluded with a special issue of the magazine devoted to Mars, in which von Braun and others described how to get there and predicted what might be found based on recent scientific data.[13]

The *Collier's* series catapulted von Braun into the public spotlight like none of his previous activities had been able to do. The magazine was one of the four highest circulation periodicals in the United States during the early 1950s, with over three million copies produced each week. If estimates of readership were indeed four or five people per copy, as the magazine claimed, something on the order of 15 million people were exposed to these space flight ideas. *Collier's,* seeing that it had a potential blockbuster, did its part by hyping the series with window advertisements of the space artwork appearing in the magazine, sending out more than 12,000 press releases, and preparing media kits. It set up interviews on radio and television for von Braun and the other space writers, but especially von Braun, whose natural charisma and enthusiasm for space flight translated well through that medium. Von Braun appeared on NBC's "Camel News Caravan" with John Cameron Swayze, on NBC's "Today" show with Dave Garroway, and on CBS's "Gary Moore" program. While *Collier's* was interested in selling magazines with these public appearances, von Braun was interested in selling the idea of space travel to the public.[14]

Following close on the heels of the *Collier's* series, Walt Disney Productions contacted von Braun—through Willy Ley—and asked his assistance in the production of three shows for Disney's weekly television series. The first of these, "Man in Space," premiered on Disney's show on March 9, 1955, with an estimated audience of 42 million. The second show, "Man and the Moon," also aired in 1955 and sported the powerful image of a wheel-like space station as a launching point for a mission to the Moon. The final show, "Mars and Beyond," premiered on December 4, 1957, after the launching of Sputnik I. Von Braun appeared in all three films to explain his concepts for human space flight, while Disney's characteristic animation illustrated the basic principles and ideas with wit and humor.[15]

While some scientists and engineers criticized von Braun for his blatant promotion of both space flight and himself, the *Collier's* series of articles and especially the three Disney television programs were exceptionally important in changing public attitudes toward space flight. Media observers noted the favorable response to the three Disney shows from the public, and recognized that "the thinking of the best scientific minds working on space projects today" went into them, "making the picture[s] more fact than fantasy."[16] Clearly the *Collier's* and Disney series helped to shape the public's perception of space flight as something that was no longer fantasy.

The coming together of public perceptions of space flight as a near-term reality with the technological developments then being seen at White Sands and elsewhere, created an environment much more conducive to the establishment of an aggressive space program. Convincing the American public that space flight was *possible* was one of the most critical components of the space policy debate of the 1950s. Without it, NASA and the aggressive piloted programs of the 1960s could never have been approved. To be approved in the public policy arena, the public must have both an appropriate vision of the phenomenon with which the society seeks to grapple and confidence in the attainability of the goal. Indeed, space enthusiasts were so successful in promoting their image of human space flight as being just over the horizon, that when other developments forced public policy makers to consider the space program seriously, alternative visions of space exploration remained ill-formed, and even advocates of different futures emphasizing robotic probes and applications satellites were obliged to discuss space exploration using the symbols of the human space travel vision that had been so well established in the minds of Americans by the promoters.[17]

The Role of Foreign Policy and National Security Issues

At the same time that space exploration advocates, both buffs and scientists, were generating an image of space flight as genuine possibility and no longer fantasy and proposing how to accomplish a far-reaching program of lunar and planetary exploration, another critical element entered the picture, the role of space flight in national defense and international relations. Space partisans early began hitching their exploration vision to the political requirements of the Cold War, in particular to the belief that the nation that occupied the "high ground" of space would dominate the territories underneath it. In the first of the *Collier's* articles in 1952, the exploration of space was framed in the context of the Cold War rivalry with the Soviet Union and concluded that "Collier's believes that the time has come for Washington to give priority of attention to the matter of space superiority. The rearmament gap between the East and West has been steadily closing. And nothing, in our opinion, should be left undone that might guarantee the peace of the world. It's as simple as that." The magazine's editors argued "that the U.S. must immediately embark on a long-range development program to secure for the West 'space superiority.' If we do not, somebody else will. That somebody else very probably would be the Soviet Union."[18]

The Cold War rivalry with the Soviet Union was the key that opened the door to aggressive space exploration, not as an end in itself, but as a means to achieving technological superiority in the eyes of the world over an adversary. From the perspective of the 1990s it is difficult to appreciate the near-hysterical concern of how nuclear attack preoccupied Americans in the 1950s. Far from being the "Happy Days" of the television sitcom, the United States was a dysfunctional nation preoccupied with death by nuclear war. Schools required children to practice civil defense techniques and shield themselves from nuclear blasts, in some cases by simply crawling under their desks. Communities practiced civil defense drills and families built personal bomb shelters in their backyards.[19] In the popular culture, nuclear attack was inexorably linked to the space above the United States, from which the attack would come.

The perception of space as the "high ground" of Cold War competition gained credibility from the atomic holocaust literature of the era.[20] In 1948 readers of *Collier's* magazine were treated to an article titled "Rocket Blitz From the Moon," which opened with a striking full-color illustration by space artist Chesley Bonestell of two V-2 shaped rockets lifting off from the lunar surface. The adjoining page showed two large fireballs spreading across an aerial view of New York City.[21] In 1947 science fiction writer Robert Heinlein teamed up with Navy Captain Caleb Laning to warn *Collier's* readers that "space travel can and will be the source of supreme military power over this planet."[22]

The danger of surprise attacks had been burned into the national consciousness by the Japanese attack on Pearl Harbor. Bonestell painted a number of nuclear holocaust pictures, including the cover for the August 5, 1950, *Collier's* magazine issue that showed an atomic blast leveling Manhattan from the point of view of an airplane approaching La Guardia airport. A similar air burst graced the April 21, 1953, issue of *Look* magazine.[23]

Couple this type of terror literature with the reality of the Soviet Union successfully testing an atomic bomb on August 29, 1949, in Semipalatinsk, Siberia, and the nightmare had become reality. This shock was still reverberating when the Soviets tested their first hydrogen bomb in the early 1950s. After an arms race that had a definite nuclear component and a series of hot and cold crises in the Eisenhower era, with the launching of Sputnik in 1957 the threat of holocaust for most Americans was now not just a possibility but a probability. One of Lyndon Johnson's aides, George E. Reedy, summarized the feelings of many Americans at that time: "the simple fact is that we can no longer consider the Russians to be behind us in technology. It took them four years to catch up to our atomic bomb and nine months to catch up to our hydrogen bomb. Now we are trying to catch up to their satellite." Then Senator John F. Kennedy agreed during the 1960 presidential campaign that "if the Soviets control space they can control earth, as in past centuries the nation that controlled the seas dominated the continents."[24]

The linkage between the idea of progress manifested through the frontier, the selling of space flight as a reality in American popular culture, and the Cold War rivalries between the United States and the Soviet Union made possible the adoption of an aggressive space program in the early 1960s. The NASA space program through Project Apollo, with its emphasis upon human space flight and extraterrestrial exploration, emerged from these three major ingredients, with Cold War concerns the dominant driver behind monetary appropriations for space efforts.

THE SPACE POLICY DEBATE IN THE 1990s

The Declining Role of the Frontier Metaphor in Public Discourse

It began to be perceptible in the late 1960s, and was certainly recognized in the 1970s, that the intermix of frontier imagery, popular culture expectations, and Cold War concerns was beginning to break down. First, the construct of the frontier as a positive image of national character and of the progress of democracy has been challenged on all quarters and virtually rejected as a useful ideal in

American postmodern, multicultural society. Western historian Patricia Nelson Limerick, for one, argued that the frontier myth, used as a happy metaphor by many, should be seen as a pejorative reflection. She argued that it denotes conquest of place and peoples, exploitation without environmental concern, wastefulness, political corruption, executive misbehavior, shoddy construction, brutal labor relations, and financial inefficiency. Limerick suggested that when the old western American frontier is conjured as an image that NASA is seeking to advance into space that someone from the space agency should punch the speaker "for insulting the organization's honor. It's a wonder no one—no shuttle pilot, mission coordinator, mechanic, or technician—said, 'Now cut that out—we may have our problems, but it's nowhere near that bad.' "[25]

Conservative politicians became the bearers of the frontier mythology increasingly used to justify the space program as the Cold War slipped away, while liberals grew increasingly restless with the exploitation and oppression that the frontier myth seemed to imply. NASA leaders have largely ignored the negative images conjured up in an increasing number of American minds by the metaphor of the frontier. For all their hard-headed practicality, for all their understanding of science and technology, they have been caught up in frontier allusion even to the present. For instance, James C. Fletcher, NASA administrator between 1971 and 1977 and again between 1986 and 1989 commented:

> History teaches us that the process of pushing back frontiers on Earth begins with exploration and discovery is followed by permanent settlements and economic development. Space will be no different... Americans have always moved toward new frontiers because we are, above all, a nation of pioneers with an insatiable urge to know the unknown. Space is no exception to that pioneering spirit.[26]

Astronaut, then senator, John Glenn captured some of this same tenor in 1983 when he summoned images of the American heritage of pioneering and argued that the next great frontier challenge was in space. "It represents the modern frontier for national adventure. Our spirit as a nation is reflected in our willingness to explore the unknown for the benefit of all humanity, and space is a prime medium in which to test our mettle."[27]

The image of the frontier, however, has been a less and less acceptable and effective metaphor near the end of the twentieth century. Liberals have come to view the space program from a quite different perspective. To the extent that space represents a new frontier, it conjures up images of commercial exploitation and the subjugation of oppressed peoples. Implemented through a large aerospace industry, in their view, it appears to create the sort of governmental-corporate complexes of which liberals are increasingly wary. Despite the promise that the Shuttle, like jet aircraft, would make space flight accessible to the "common man," space travel remains the province of a favored few, perpetuating inequalities rather than leveling differences. They also assert that space exploration has also remained largely a male frontier, with room for few minorities. In the eyes of liberals, space perpetuates the inequities that they have increasingly sought to abolish on Earth. As a consequence, it is not viewed favorably by those caught up in what political scientist Aaron Wildavsky has characterized as "the rise of radical egalitarianism."[28] The advent of this new liberal philosophy coincides with the shift in ideological positions on the U.S. space program in the late 1960s and early 1970s.

The Widening Gap Between Popular Conceptions of Space Flight and Its Reality

Second, the narrowing of the gap between what the public perceived as the probability of space flight and its reality in the 1950s and 1960s widened again in the 1980s and 1990s. This was largely because NASA's and others' promotional efforts did not square with what actually took place. When the space promoters of the 1950s described trips to the Moon, the reality of their accomplishment reinforced the promoters' images. Officials of NASA and other advocates of aggressive space exploration goals addressed the dwindling support that results from this breakdown in a number of ways. Most important, they sought to reduce the cost of space flight, hoping to make space more accessible, an excellent shift in approach that was full of promise. In 1970, the President's Science Advisory Committee accurately pointed out that the future of human space flight would be determined by the outcome of efforts to control flight costs.

The Space Shuttle, on the drawing board since the 1960s, but a major program in its own right only after 1970, was an attempt to provide "low-cost access [to space] by reusable chemical and nuclear rocket transportation systems."[29] George M. Low, NASA's deputy administrator, voiced the redefinition of this approach to the NASA leadership on January 27, 1970: "I think there is really only one objective for the Space Shuttle program, and that is 'to provide a low-cost, economical space transportation system.' To meet this objective, one has to concentrate both on low development costs and on low operational costs."[30] "Low cost, economical" space transportation became NASA's criteria for the program, and it was an effort to deal with a real-time problem of public perception about space flight at the time: that it was too expensive.[31]

While the goal of "low cost, economical" access to space was a useful tool for meeting public concerns about the possibilities of space flight, it eventually proved an embarrassment to the space program. So far, costs of space flight have not declined appreciably. As a result when the Shuttle, advertized as the world's first spaceliner, failed to achieve ready, economical access to space, the previously close connection between reality in space exploration and the public's perceptions of it sharply diverged. This was enormously exacerbated by the *Challenger* accident in January 1986, which graphically demonstrated before the world that the Shuttle was much more of a test and far less of an operational vehicle than NASA leaders had publicly said.[32]

Third, just as the Cold War was the driving force behind big budgets in the 1960s, its end has been a critical component in the search for a new space policy in the 1990s. It has also reinforced the quest for low cost, economical space exploration. NASA tried to build support for its programs on the basis of competition between the United States and the Soviet Union as long as possible. James E. Webb criticized the Johnson administration for reducing NASA's fiscal year 1970 budget on the ground that it ensured that the United States would fall behind the Soviet Union in space and national prestige.[33] James Fletcher appealed to Nixon to approve the building of the Shuttle because, among other things, it might be useful in capturing and recovering both American and Soviet satellites.[34] On the other hand, cold war rivalries no longer held any attraction as a selling point for an aggressive space program at least by the mid-1980s.

A New Turn on the Foreign Affairs and National Security Issue

What did offer promise, however, has been a different spin on international relations, that is, cooperation with European partners. While Apollo had been a totally American project because of its Cold War origins to best the Soviet Union, in post-Apollo activities there were opportunities for greater partnership with foreign allies.

NASA flirted with the predecessor of the European Space Agency in the early 1970s over involvement with the Shuttle, and eventually European partners built Spacelab for use in the vehicle.[35]

From the outset of the space station program of the 1980s, both NASA and the Reagan administration held it to be an international effort. NASA forged international agreements among thirteen nations to take part in the Space Station *Freedom* program. Japan, Canada, and the nations pooling their resources in the European Space Agency (ESA) agreed in the spring of 1985 to participate. It was thought that this would help maximize technological capability while reducing the cost to each partner. But the really ingenious aspect to the partnership arrangement was that it helped to replace the rivalry of the Cold War as a means of stabilizing support and funding for NASA by tying the space program's major program to yet another U.S. foreign policy objective, in this case international cooperation. In so doing, every partnership brought greater legitimacy to the overall program and helped to insulate it from drastic budgetary and political changes. Inciting an international incident because of a change to the station was something neither U.S. diplomats nor politicians relished, and that fact, it was thought, could help stabilize funding, schedule, or other factors that might otherwise be changed in response to short-term political needs.[36]

This new spin on an old idea—linking the space program to international relations—took an even more significant turn in the fall of 1993 when NASA negotiated a landmark decision to include Russia in the building of an international space station. In the post–Cold War era, as the United States is wrestling with foreign policy questions that are aimed at supporting democratic reforms in eastern Europe and Russia, this decision has thus far provided an important linkage for the continuation of the space station effort at a time when too many politicians are convinced that for all other reasons—cost, technological challenge, return on investment, etc.—it is an uninviting endeavor. Like the members of Congress of the early 1960s who were disinterested in space exploration except as a tool of foreign policy goals, the modern cooperative agreements for joint space exploration seem to prove more important to some members than the other issues associated with the effort.

CONCLUSION

The combination of technological and scientific advance, political competition with the Soviet Union, and changes in popular opinion about space flight came together in a very specific way in the 1950s to affect public policy in favor of an aggressive space program. This found tangible expression in efforts in the 1950s and 1960s to move forward with an expansive space program and the budgets necessary to support it.

Likewise, the central ingredients of space policy formulation in the 1990s have altered appreciably or disintegrated altogether. The frontier metaphor as a useful tool of advocating space exploration is gone. Only conservative politicians and advocates still invoke it, and their opponents find the argument not only worthless but derogatory. The gap between reality and popular ideas of space flight is so wide that a new campaign will be required to link the two once again. That might well happen, but the development of a new technology that will be able to deliver on the promise of space flight for humanity to leave this planet will need to be demonstrated first. Finally, the coupling of the space effort to international relations has undergone a transition from rivalry to cooperation, and while this may be a temporary condition, at least for the present it seems to be the overarching force behind the maintenance of a political coalition providing funding to the space station, the most expensive program of NASA that has been advocated over the years.

NOTES

1. See, Ray A. Williamson, "Outer Space as Frontier: Lessons for Today," *Western Folklore,* 46 (October 1987): 255–67; Stephen J. Pyne, "Space: A Third Great Age of Discovery," *Space Policy,* 4 (August 1988): 187–99; John Glenn, Jr., "The Next 25: Agenda for the U.S.," *IEEE Spectrum,* September 1983, p. 91; James A. Michener, "Looking Toward Space," *Omni,* May 1980, pp. 57–58, 121; James A. Michener, "Manifest Destiny," *Omni,* April 1981, pp. 48–50, 102–104; G. Harry Stine, *The Hopeful Future* (New York: The Macmillan Co., 1983); *America's Next Decades in Space: A Report of the Space Task Group* (Washington, DC: National Aeronautics and Space Administration, September 1969); Harvey Brooks, "Motivations for the Space Program: Past and Future," In Allan A. Needell, ed., *The First 25 Years in Space: A Symposium* (Washington, DC: Smithsonian Institution Press, 1983), pp. 3–26.

2. On this subject see, Richard E. Neustadt and Ernest R. May, *Thinking in Time: The Uses of History for Decision Makers* (New York: The Free Press, 1986). On the limitations of analogy in historical study see Bruce Mazlish, ed., *The Railroad and the Space*

Program: An Exploration in Historical Analogy (Cambridge, MA: Harvard University Press, 1965).

3. Neustadt and May, *Thinking in Time,* pp. 273–75.

4. On these factors see, Michael A. G. Michaud, *Reaching for the High Frontier: The American Pro-Space Movement, 1972–1984* (New York: Praeger, 1986); Emmette Redford and Orion F. White, *What Manned Space Program After Reaching the Moon? Government Attempts to Decide, 1962–1968* (Syracuse, NY: The Inter-University Case Program, January 1971); Klaus P. Heiss and Oskar Morgenstern, *Mathematica Economic Analysis of the Space Shuttle System* (Princeton, NJ: Mathematica, Inc., 1972); Ken Hechler, *Toward the Endless Frontier: History of the Committee on Science and Technology, 1959–1979* (Washington, DC: U.S. House of Representatives, 1980); *The Next Decade in Space: A Report of the Space Science and Technology Panel of the President's Science Advisory Committee* (Washington, DC: President's Science Advisory Committee, March 1970); President's Science Advisory Committee, "A Statement by the President and the Introduction to Outer Space," 26 March 1958, NASA Historical Reference Collection, NASA Headquarters, Washington, DC; Radford Byerly, Jr., ed., *Space Policy Reconsidered* (Boulder, CO: Westview Press, 1989); Radford Byerly, Jr., ed., *Space Policy Alternatives* (Boulder, CO: Westview Press, 1993); Richard P. Hallion and Tom D. Crouch, eds. *Apollo: Ten Years Since Tranquility Base* (Washington, DC: Smithsonian Institution Press, 1979); John M. Logsdon, et al., *Apollo in Its Historical Context* (Washington, DC: George Washington University Space Policy Institute, 1990); John R. Pottenger, "Space, Science, and Society: Technology and the Rational State," unpublished manuscript, NASA Historical Reference Collection.

5. This is an expression of Frederick Jackson Turner's "Frontier Thesis" that guided inquiry into much of American history for a generation. It also continues to inform many popular images of the American West. Turner outlined the major features of the subject in Frederick Jackson Turner, *The Frontier in American History* (New York: Holt, Rinehart, and Winston, 1920; Reprint, Malabar, FL: Krieger Publishing Co., 1985), which included the seminal 1893 essay, "The Significance of the Frontier in American History."

6. This frontier imagery was overtly mythic. Myths, however, are important to the maintenance of any society, for they are stories that symbolize an overarching ideology and moral consciousness. As James Oliver Robertson observes in *American Myth, American Reality* (1980), "Myths are the patterns of behavior, or belief, and/or

perception—which people have in common. Myths are not deliberately, or necessarily consciously, fictitious" (James Oliver Robertson, *American Myth, American Reality* (New York: Hill and Wang, 1980), p. xv). Myth, therefore, is not so much a fable or falsehood, as it is a story, a kind of poetry, about events and situations that have great significance for the people involved. Myths are, in fact, essential truths for the members of a cultural group who hold them, enact them, or perceive them. They are sometimes expressed in narratives, but in literate societies like the United States they are also apt to be embedded in ideologies. Robertson's book is one of many studies that focus on American myths—such as the myth of the chosen people, the myth of a God-given destiny, and the myth of a New World innocence or inherent virtue.

7. This is the thesis of William Sims Bainbridge, *The Spaceflight Revolution: A Sociological Study* (New York: Wiley, 1976). See also Willy Ley and Chesley Bonestell, *The Conquest of Space* (New York: Viking, 1949).

8. George H. Gallup, *The Gallup Poll: Public Opinion, 1935–1971* (New York: Random House, 1972), 1:875, 1152.

9. One of the keys in this process was the work of film producer-director George Pal, a master of special effects, who made several space-oriented movies in the 1950s. On Pal's career see, Gail Morgan Hickman, *The Films of George Pal* (South Berwick: A.S. Barnes, 1977); Robert A. Heinlein, "Shooting Destination Moon," *Astounding Science Fiction,* July 1950, p. 6. Especially memorable were two films, *The Day the Earth Stood Still* (1950), directed by Robert Wise, in which the benevolent alien Klaatu warns the Earth to shape up and control its aggressiveness by disarming, and *Forbidden Planet* (1956), about the extinct Krell superintelligent society and the Monster from the Id. These films excited the public with ideas of space flight, exploration, and contact with alien civilizations. It is often easy to forget that these sophisticated visions of space travel occurred before Sputnik.

10. For example, science fiction writer Arthur C. Clarke described in February 1945 the use of the German V-2 as a launcher for ionospheric research, even as the war was going on. He specifically suggested that by putting a second stage on a V-2 the rocket could generate enough velocity to launch a small satellite into orbit. "Both of these developments demand nothing in the way of technical resources," he wrote, adding that they "should come within the next five or ten years." See, Arthur C. Clarke, "V2 for Ionospheric Research?" *Wireless World,* February 1945, p. 58.

11. See as an example of his exceptionally sophisticated space flight promoting, Wernher von Braun, *The Mars Project* (Urbana: University of Illinois Press, 1953), based on a German-language series of articles appearing in the magazine *Weltraumfahrt* in 1952.

12. "Man Will Conquer Space Soon" series, *Collier's,* March 22, 1952, pp. 23–76ff.

13. Wernher von Braun with Cornelius Ryan, "Can We Get to Mars?" *Collier's,* April 30, 1954, pp. 22–28.

14. Randy L. Liebermann, "The *Collier's* and Disney Series," in Frederick I. Ordway III and Randy L. Liebermann, *Blueprint for Space* (Washington, DC: Smithsonian Institution Press, 1992), p. 141; Ron Miller, "Days of Future Past," *Omni,* October 1986, pp. 76–81.

15. Liebermann, "The *Collier's* and Disney Series," in Ordway and Liebermann, *Blueprint for Space,* pp. 144–46; David R. Smith, "They're Following Our Script: Walt Disney's Trip to Tomorrowland," *Future,* May 1978, pp. 59–60; Mike Wright, "The Disney–Von Braun Collaboration and Its Influence on Space Exploration," paper presented at conference, "Inner Space, Outer Space: Humanities, Technology, and the Postmodern World," 12–14 February 1993; Willy Ley, *Rockets, Missiles, and Space Travel* (New York: The Viking Press, 1961 ed.), p. 331.

16. *TV Guide,* March 5, 1955, p. 9.

17. The dichotomy of visions has been one of the central components of the U.S. space program. Those who advocated a scientifically oriented program using nonpiloted probes and applications satellites for weather, communications, and a host of other useful activities were never able to capture the imagination of the American public the way the human spaceflight advocates did. For a modern critique of this dichotomy see, Alex Roland, "Barnstorming in Space: The Rise and Fall of the Romantic Era of Spaceflight, 1957–1986," in Byerly, ed., *Space Policy Reconsidered,* pp. 33–52. That the human imperative is still consequential is demonstrated in William Sims Bainbridge's sociological study, *Goals in Space: American Values and the Future of Technology* (Albany: State University of New York Press, 1991).

18. "What Are We Waiting For?" *Collier's,* March 22, 1952, p. 23.

19. See, Elaine Tyler May, *Homeward Bound: American Families in the Cold War Era* (New York: Basic Books, 1988), pp. 93–94, 104–13.

20. See, "The A-Bomb's Invisible Offspring," *Collier's,* August 9, 1947; "Power on Glory and Wings," *Collier's,* March 27, 1948; "Fear,

War and the Bomb," *New Republic,* November 29, 1948; "Defense Against Atom-Bomb Blitz? None Yet," *Newsweek,* November 14, 1949. Once of these includes a statement that then-Congressman John F. Kennedy was one of the few public figures concerned about the dangers of atomic attack.

21. Robert S. Richardson, "Rocket Blitz from the Moon," *Collier's,* October 23, 1948, pp. 24–25.

22. "The A-Bomb's Invisible Offspring," *Collier's,* August 9, 1947.

23. Ron Miller and Frederick C. Durant, *Worlds Beyond: The Art of Chesley Bonestell* (Norfolk, VA: Donning, 1983); John Lear, "Hiroshima, U.S.A.," *Collier's,* August 5, 1950, cover; William L. Laurence, "How Hellish Is the H-Bomb?" *Look,* April 21, 1953, p. 31.

24. Lee D. Saegesser, "High Ground Advantage," NASA Historical Reference Collection. Also see John F. Kennedy, "If the Soviets Control Space," *Missiles and Rockets,* October 10, 1960, p. 12.

25. Patricia Nelson Limerick, "The Final Frontier?" *Wilson Quarterly,* 14 (Summer 1990): 82–83, quote from p. 83.

26. James C. Fletcher, "Our Space Program Is Already Back on Track," *USA Today,* 28 July 1987.

27. Glenn, "The Next 25: Agenda for the U.S.," p. 91.

28. Aaron Wildavsky, *The Rise of Radical Egalitarianism* (Washington, DC: American University Press, 1991).

29. NASA, *The Post-Apollo Space Program: A Report for the Space Task Group* (Washington, DC: National Aeronautics and Space Administration, September 1969), p. 6.

30. George M. Low to Dale D. Myers, "Space Shuttle Objectives," 27 January 1970, NASA History Division Reference Collection.

31. In January 1970 Thomas O. Paine, Richard Nixon's appointee as the NASA administrator, described a somber meeting with the president in which Nixon told him that both public opinion polls and political advisors indicated that the mood of the country suggested hard cuts in the space and defense programs. Memo by Thomas O. Paine, "Meeting with the President, January 22, 1970," January 22, 1970, NASA Historical Reference Collection: Caspar W. Weinberger interview by John M. Logsdon, August 23, 1977, NASA Historical Reference Collection.

32. On this issue see Robert W. Kubey and Thea Peluso, "Emotional Response as a Cause of Interpersonal News Diffusion: The Case of the Space Shuttle Tragedy," *Journal of Broadcasting & Electronic Media,* 34 (Winter 1990): 69–76; Jon D. Miller, "The *Challenger* Accident and Public Opinion: Attitudes Toward the Space Programme in the USA," *Space Policy,* 3 (May 1987): 122–40; Daniel

Riffe and James Glen Stovall, "Diffusion of News of Shuttle Disaster: What Role for Emotional Response," *Journalism Quarterly,* 66 (Autumn 1989): 551–56; Alex Roland, "Priorities in Space for the USA," *Space Policy,* 3 (May 1987): 104–14; Gregory Whitehead, "The Forensic Theater: Memory Plays for the Post-Mortem Condition," *Performing Arts Journal,* 12 (Winter-Spring 1990): 99–110; John C. Wright, Dale Kunkel, Marites Pinon, and Aletha C. Houston, "How Children Reacted to Televised Coverage of the Space Shuttle Disaster," *Journal of Communication,* 39 (Spring 1989): 27–45.

33. Donald F. Horning to Lyndon B. Johnson, 26 September 1968; Lyndon B. Johnson to Donald F. Hornig, 26 September 1968, both in Executive Office Series, White House Central Files, Johnson Library.

34. Jacob E. Smart, NASA assistant administrator for DOD and Interagency Affairs, to James C. Fletcher, NASA administrator, "Security Implications in National Space Program," 1 December 1971, with attachments, James C. Fletcher Papers, Special Collections, Marriott Library, University of Utah, Salt Lake City, UT; James C. Fletcher, NASA administrator, to George M. Low, NASA deputy administrator, "Conversation with Al Haig," 2 December 1971, NASA History Division Reference Collection.

35. Thomas O. Paine to the president, 7 November 1969, White House, Richard M. Nixon, president, 1968–1971 File, NASA History Office Reference Collection; Howard E. McCurdy, *The Space Station Decision: Incremental Politics and Technological Choice* (Baltimore, MD: The Johns Hopkins University Press, 1990), p. 101; David Shapland and Michael Rycroft, *Spacelab: Research in Earth Orbit* (Cambridge, England: Cambridge University Press, 1984); Douglas R. Lord, *Spacelab: An International Success Story* (Washington, DC: National Aeronautics and Space Administration, 1987).

36. John M. Logsdon, "International Cooperation in the Space Station Programme: Assessing the Experience to Date," *Space Policy,* 7 (February 1991), 38, 41–44; John M. Logsdon, *Together in Orbit: The Origins of International Participation in Space Station Freedom* (Washington, DC: George Washington University, 1991), pp. 37, 129–130; Kenneth S. Pedersen, "Thoughts on International Space Cooperation and Interests in the Post-Cold War World," *Space Policy,* 8 (August 1992), 217.

11
COLOR MATTERS: RACE, HISTORY, AND FLORIDA'S SARA LEE DOLL
Gordon Patterson

"It had begun with Christmas," declared Claudia, Toni Morrison's narrator in her novel *The Bluest Eye*. "The big, the special, the loving gift was always a big, blue-eyed, Baby Doll. From the clucking sounds of adults I knew that the doll represented what they thought was my fondest wish. I was bemused with the thing itself, and the way it looked. What was I supposed to do with it?"[1] Through much of the twentieth century, generations of African-American children experienced the full force of this fictional character's dilemma. Their skin color posed a barrier between their self-image and the image of innumerable blue-eyed, yellow-haired, pink-skinned dolls. In Morrison's novel, Claudia reacted by declaring that she did not love babies and did not want to grow up and be a mother. Later, she struck out. She destroyed the white dolls. "But the dismembering of dolls," she relates, "was not the true horror. The truly horrifying thing was the transference of the same impulses to little white girls. The indifference with which I could have axed them was shaken only by my desire to do so. To discover what eluded me: the secret of the magic they weaved on others. What made people look at them and say, 'Awwwww,' but not for me?"[2]

Twenty-two years before Morrison wrote her fictional account of a black girl's childhood, Sara Lee Creech, a white woman, stopped at the post office in Belle Glade, Florida, one day in December 1948. Returning to her car, Creech noticed two black children playing dolls in the back seat of a Buick. The little girls were playing with white dolls. It was wrong, she thought, that black children did not have

---

Reprinted with permission from "Color Matters: The Creation of the Sara Lee Doll," *Florida Historical Quarterly* 73, (October 1994), 147–65.

quality colored dolls to play with. Later she phoned her friend Maxeda von Hesse in New York and asked if she would help create an African-American doll that would represent the beauty and diversity of black children.[3]

Neither woman knew anything about dolls nor the toy industry. Creech supported herself through a flower shop and an insurance agency. Nevertheless, in 1948 the women set out to create an "anthropologically correct" black doll.[4] Three years later Eleanor Roosevelt gave a reception for their creation, the Sara Lee Doll. Ralph Bunche, Walter White, Jackie Robinson, Winthrop Rockefeller, David Rockefeller, and Bernard Baruch were among the guests.[5] Sears and Roebuck introduced the Sara Lee Doll in its 1951 Christmas catalogue. "This Christmas," a reporter wrote in the January 1952 edition of *Ebony* magazine, "a half million little girls of many races found under their Christmas trees some of the most beautiful Negro dolls America has ever produced. A transformation has taken place in toyland and new colored dolls with delicate features, lighter skin, and modish clothes are being introduced in the world of childhood fantasy where always before the Negro doll was presented as a ridiculous, calico-garmented, handkerchief-headed servant."[6]

Issues of race, color, commercialism, and popular culture converge in the story of the Sara Lee Doll, which grew out of a woman's desire to forge an interracial alliance. In 1949 Creech enlisted the support of prominent African Americans and white liberals. These reformers shared a commitment to fighting prejudice. The Sara Lee Doll's supporters worked for two years to create positive, nonstereotyped toys for both black and white children. In 1951 the Ideal Toy Company manufactured the Sara Lee Doll. Ultimately, however, color and commercialism proved insurmountable obstacles to realizing their objective.

Sara Creech began to work in the women's and interracial movements in the mid 1930s in Lake Worth. In 1941 she moved inland to Belle Glade. While in Lake Worth, Creech had joined the local Business and Professional Women's Club (BPWC). Shortly after her arrival in the Glades, she organized a new BPWC chapter. In 1946 she was elected second vice president of the state BPWC, and a year later she was elected president of the state federation of Business and Professional Women's Clubs.

Creech's involvement in the BPWC put her in contact with Edna Giles Fuller, a past leader of the association, who had founded Florida's first Inter-Racial Council in Orlando. Creech believed that Belle Glade needed such a group. Since 1944 four men (two whites and two blacks) in Belle Glade had met informally to discuss community issues. In 1947 these men asked Creech to join them. She suggested that they pattern their efforts after Fuller's model. On April 4, 1948, the Belle Glade Inter-Racial Council held its first regular meeting.

Creech's work on the Inter-Racial Council soon eclipsed her BPWC activities. Daily contact with the black and migrant communities strengthened her conviction that change was necessary. Louise Taylor, wife of one of the council's founders, was Creech's friend. In the weeks before Christmas 1948, Taylor described to Creech the challenges she faced raising her three children. "I had never thought about it," Creech remembered. "I had never thought about a black child needing a black doll to play with."[7]

The sight of the two black girls playing with white dolls in the back of the Buick was Creech's epiphany. That evening Creech did two things. She told her mother about her idea of making a quality black doll. Her mother responded: "Sara, you know absolutely nothing about manufacturing. I don't see how you can finance it. But I can see the importance of it. If you want to try it, I'll back you in whatever you do."[8] It was then that Creech called her friend Maxeda von Hesse in New York and succeeded in gaining her support. Von Hesse would find an artist to sculpt the doll's head; Creech would use her contacts to develop the doll's concept.[9]

Over the next two years Creech became an expert on the history of colored dolls. The 1950 patent application for the Sara Lee Doll documents her study. Creech's research demonstrated that the typical nineteenth- and twentieth-century African-American doll was a stereotyped Mammy or Pickaninny doll that presented African Americans as objects of comedy or ridicule. Creech believed that race prejudice was transmitted to children through dolls and games. In 1929 sociologist Bruno Lasker cited an article in the *Children's Encyclopedia* entitled "Favorite Garden Games" as evidence of the "negative educational by-product" of such playthings. The game in question was called Aunt Sally. "Aunt Sally is a black doll. She wears a white cap on her head and a white cape on her shoulders, and carries a pipe loosely in her mouth. Her body is only a stick with a pointed end, and when this is pushed into the ground she is ready for the fun to begin. The players stand at a distance of some yards and, each in turn, throw at the pipe with a number of short, stout sticks. Those who knock it out of her mouth the greatest number of times win the game, but those who cannot aim straight must not be surprised if Aunt Sally seems to smile at them."[10]

Creech learned that there had been earlier attempts to provide alternatives to Aunt Sally and other "coon dancer[s]" dolls.[11] At the turn of the century some black church leaders, recognizing the educational role of dolls, resolved to break the stereotypes and market quality dolls. Dr. R. H. Boyd, a leader of the Nashville-based National Baptist Publishing House, traveled to Europe to convince German toy manufacturers to produce a nonstereotyped black doll. The toy makers initially resisted Boyd's proposal. Apparently, they believed that there was no market for the realistic doll. Boyd persisted, sending scores of photographs of African Americans to the companies. Eventually, Boyd succeeded in getting the doll manufactured, and they were sold in black churches until the beginning of World War I.[12] After the war, an African-American company, the leaders of whom followed Marcus Garvey, resumed importing the German bisque dolls. They hoped to use black dolls to develop children's race consciousness.[13]

Presumably, most American toy companies were not interested in producing realistic black dolls. The few companies that did produce colored dolls followed the P & M Doll Company's lead and painted white dolls "chocolate brown."[14] P & M's success with its colored Daisy Doll led the company to introduce the Topsy doll. "Topsy," however, was criticized because of its stereotypical "black face, banjo eyes, and three little pigtails."[15] Simultaneously, Aunt Jemima Mills marketed a line of stereotyped dolls based on Aunt Jemima and Uncle Mose.[16] Thus, until the 1930s most American toy manufacturers produced colored dolls that caricatured African Americans as "mammies" and "coon dancers."[17] In 1931 the Allied-Grand Doll Manufacturing Company expanded the procedure of painting white dolls brown and advertising them as "Negroid."[18]

World War II had a tremendous impact on the toy industry. The end of the war marked the beginning of the baby boom. Toy manufacturers expanded production to keep pace with the new families. Returning African-American servicemen and women resisted Jim Crowism in the workplace and the playground. In 1948 the Terri Lee Corporation introduced its teenaged Patti-Jo Doll, which cost $15.95. The light-skinned Patti-Jo and her boyfriend, Benji, were among the first new offerings. A year later the Sun Rubber Company marketed an Amosandra doll. It was a spin-off of the popular Amos 'n Andy radio show. And in 1950 Allied Grand introduced a Jackie Robinson doll.

Sara Creech's research into these early dolls led her to the conclusion that parents who wished to purchase a colored baby doll for their children had two choices: They could buy either a grotesquely stereotyped doll or a white doll that had been shaded brown. She did not want to produce a teenage, glamorous doll like Patti-Jo or to pattern her doll after a national figure like Jackie Robinson. Rather she wanted to create a doll that would "reflect their [African-American children's] *own* attractiveness." She reasoned that "in the game of Make-Believe-Grown-Ups, how much more normal and healthy would be their play if they learned from *their* dolls a wholesome self-respect and appreciation for their own heritage."[19] Like R. H. Boyd a half-century earlier, Creech wanted her doll to bear a realistic likeness to American blacks. Her goal was to create dolls that would capture "the simplicity and natural dignity of the finest type of our colored children."[20]

By early January 1949 the doll project was taking shape. Maxeda von Hesse had enlisted Sheila Burlingame to join the project and sculpt the heads. She was the sculptor of the *Negro Boy Praying*, which the St. Louis Urban League had placed before its headquarters.

Burlingame outlined her requirements: she needed pictures of African-American children and head measurements. Creech accumulated more than 500 pictures, and Burlingame selected a handful of pictures in order to prepare sketches. Creech and von Hesse needed to find a toy company willing to take a risk on the doll.[21]

While in New York City in August 1949 at the invitation of Maxeda von Hesse and her mother, the von Hesses introduced Creech to Eleanor Roosevelt. The elder von Hesse had served as Eleanor Roosevelt's speech coach. They invited Creech to a reception honoring Mrs. Roosevelt. Creech also learned that her friend, Zora Neale Hurston, was in New York, and she told the writer about the doll. Hurston liked the idea and promised to work on the project when she returned to Florida. Hurston planned to move to Belle Glade.[22] Creech knew that the task of persuading a manufacturer to produce the doll would be easier if she could show that the doll had support in the black community.

When Creech returned to Belle Glade in September 1949, she had two goals: to obtain a copyright for the doll and to locate a willing toy manufacturer. Establishing legal protection for the doll proved easy. Finding a toy company willing to commit itself to manufacturing the doll proved more difficult.[23]

Hurston played an important role in rallying support for the project. She arrived in Belle Glade the next spring, and she and Creech met virtually every evening at Creech's home. Creech remembers that during this period she began to doubt that the doll would ever become a reality. One day while she and Hurston were painting Creech's house, she told Hurston about her misgivings. "Sara," Hurston probed, "have you thought this over? Have you given it your full attention? Do you think that you are right in what you are doing?" Creech answered, "Zora, from everything I can lay my hands on, I believe a quality doll should be produced." Hurston declared, "Well, go ahead. Don't go ring'in no backin' bells."[24]

Hurston planned a strategy for winning support in the black community. While Creech focused her efforts on getting the doll manufactured, Hurston thought it essential that African-American leaders contribute to the project. She suppled Creech with letters of introduction to individuals in Atlanta, Washington, and New York.

Creech left Belle Glade in June 1950; her first stop was Atlanta. She had appointments with Bishop R. R. Wright, Jr., leader of the African Methodist Episcopal Church and president of Morris Brown College; President Benjamin Mays of Morehouse College; and President Rufus Clement of Atlanta University. A meeting was arranged with Harley Kimmel, the southern merchandise manager for Sears and Roebuck, who had access to leading toy manufacturers.

Kimmel liked the project and promised to pass the idea up the Sears chain of command. Dr. Mays wrote directly to General Robert Wood, chief executive officer at Sears, encouraging him to back the project.[25] Bishop Wright recalled his efforts fifty years earlier when he had helped R. H. Boyd import dolls from Germany. He believed that the country was ready for a quality colored doll. Mays and Bishop Wright also endorsed the project, and President Clement recommended that Creech discuss the proposal with Professor Helen Whiting, who taught early child education at Atlanta University.[26]

Whiting talked about skin color; she believed that the doll's acceptance in black families hinged on the color issue. She cautioned that the doll should not be too dark. An acceptable skin color, one that did not validate any shade as "true" or "desirable" for African Americans, was needed.[27]

Creech had not recognized the importance of the color issue. Her idea was to manufacture a quality doll and to ensure that it reflected black children as they were. Whiting raised an issue that Creech had overlooked. No single color was representative for all blacks. If the dolls were to depict what Hurston called "the beauty and character, the good features of a black child," then they must affirm the diversity of African Americans.[28] Creech believed that the solution to the problem was to create a family of dolls that displayed the diversity of the African-American community. There could be a baby doll, brother doll, sister doll, and a little Miss doll. Each would represent a different skin color and hair type. This would protect the dolls from the charge that they represented a stereotyped view of African Americans. In November 1950 Benjamin Mays reinforced Creech's belief in the need for more than one doll. He wrote that the roots of prejudice resided in what "we learn unconsciously." He reminded Creech that it was crucial that the doll not simply replace old stereotypes with new prejudices. He also suggested that all four models be put on the market simultaneously.[29]

Sara Creech received an encouraging letter from Hurston. "The thing that pleased me the most, Miss Creech," Hurston wrote, "was that you a White girl, should have seen into our hearts so clearly, and sought to meet our longing for understanding of us as we really are, and not as some would have us."[30] Hurston told Creech to contact Georgia Douglas Johnson, the black poet, and President Mordecai Johnson of Howard University. Through Georgia Douglas Johnson, Creech met many black intellectuals.[31] Winning Mordecai Johnson's support was crucial, as he was a leader in Washington's black community. He arranged for Creech to meet Lois Jones, professor of design in Howard's art department. Jones made charcoal sketches of each head and four water colors of the sister doll. She made only minor changes to Burlingame's original conception but recommended changing the hair styles—adding braids to the sister doll—and modifying the facial expressions slightly.

In New York during July and August 1950 Creech and von Hesse rallied support for the doll. James Hubert, a leader in the Urban League, gave them guidance in learning about the toy industry. He arranged a tour of the Ideal Toy Company's Jamaica, New York, plant for Creech and von Hesse.[32]

The summer had produced impressive results. The doll was under copyright, and important black leaders had indicated support for the project. The most important backer, however, was Eleanor Roosevelt. The former First Lady invited Creech and von Hesse to visit her at her Val-Kill cottage. They used the August 1950 meeting to outline their plans for the doll. "I was so interested to see the Negro dolls you are proposing to manufacture," Mrs. Roosevelt wrote after their interview. "I like them particularly because they can be made and sold on an equal basis with white dolls. There is nothing to be ashamed of. They are attractive and reproduced well with careful study of the anthropological background of the race. I think they are a lesson in equality for little children and we will find that many a child will cherish a charming black doll as easily as it will a charming white doll."[33] Mrs. Roosevelt wrote a letter of introduction for Creech and von Hesse to Ralph Bunche requesting that he advise them on the doll project.[34]

Creech returned to Belle Glade in September. There was nothing to do but wait. Harley Kimmel promised to keep her informed about his efforts to find a toy manufacturer. Maxeda von Hesse used a fall vacation to come to Belle Glade. While they were meeting, Creech received a letter from Kimmel. He wrote that he had succeeded in interesting one of Sears's purchasing directors in the doll. He asked Creech to meet in New York in mid January 1951 with Lothar Kiesow, from Sears, and David Rosenstein, president of the Ideal Toy Company.[35]

Ideal was the largest manufacturer of dolls in the world at this time. Founded at the turn of the century by Morris Michtom, creator of the Teddy Bear, Ideal had remained a family-run business. Benjamin Michtom, Morris Michtom's son, and Abe Katz, the founder's nephew, controlled the company. David Rosenstein served as company president. Rosenstein became the doll's champion at Ideal. He had a long record of involvement in social issues. Michtom and Katz were skeptical of the proposed doll and felt that Rosenstein was letting his social conscience cloud his business sense.[36]

Creech returned to Belle Glade in high spirits after her meeting with Rosenstein. On February 4, 1951, she and Hurston reported on the project to members of the Belle Glade Inter-Racial Council. She also informed the group that she had invited Mordecai Johnson to come and address the council. Johnson spoke at the Methodist church, and afterwards he and Creech discussed the doll project.[37]

When Johnson returned to Washington, he wrote to David Rosenstein and outlined the reasons why manufacturing the doll was good business. "It is my opinion that his project is timely, that it is an admirable conception, and that it is capable of wide and rewarding commercial success," he wrote.[38] Johnson underscored the significance he attached to the project. He believed that the doll could be a powerful educational tool. Black Americans were a neglected group of consumers. If Ideal took the lead in the campaign against prejudice in toys, the company would "win many, many new customer friends." This was precisely what company officials needed to hear.

Johnson realized that the campaign against racial discrimination was at a turning point. Lawyers from the NAACP's Legal Defense Fund were preparing for a trial in Clarendon County, South Carolina. This case would lead to the Supreme Court's decision three years later in *Brown* v. *Board of Education*. Dolls played a crucial role in this process. Johnson knew that Kenneth and Maime Clark had used dolls in their research into black children's self-esteem. In 1939 the Clarks had purchased four dolls; all were identical except that two were colored brown. The Clarks' findings indicated that by the time African-American children reached nursery school, they had internalized many of society's prejudices about race. In a series of articles, the Clarks reported that children viewed colored dolls as inferior simply because they were colored.[39]

Robert Carter and Thurgood Marshall called Kenneth Clark as an expert witness on the effects of segregation on black children in the Claredon County case. He testified that the children that he had examined demonstrated "an unmistakable preference for the white doll and a rejection of the brown doll." By as early as three years of age black children "suffered from self-rejection" brought on by a "corrosive awareness of color."[40]

Like many educators, Creech believed that the battle against prejudice and self-hatred had to begin in the nursery. With only white dolls and playthings, black children were more likely than whites to feel inferior. Creech considered her doll a progressive step. She believed that quality dolls, based on realistic portrayals of black children, would improve the image of African Americans for both whites and blacks.

President Charles Johnson of Fisk University supported this conclusion. He informed Creech that he "liked the general idea [of the dolls] and thought it would be an excellent thing to have them put into production." Like Benjamin Mays, Johnson advised that the four dolls be manufactured simultaneously. This would offer consumers a wider choice, and it would "remove any doubts which might arise regarding the possibility of creating a stereotype of Negro characteristics."[41]

Charles Johnson asked Beryl Shelton, director of the nursery school at Fisk, to "show the [sketches of the] dolls to a group of the children and some of the adults working with the children, and to compile the record of spontaneous reactions." Shelton showed twenty children (sixteen black and four white) the portfolio of pictures that Creech had developed. The children were enthusiastic. "No mention," Shelton noted, "was made of the fact that these were Negro dolls."[42]

The adults in Shelton's study identified the issue to which Charles Johnson had called attention in his letter to Creech. A single black doll contained the potential of becoming a "stereotype of Negro characteristics."[43] That is why Johnson advised that the four dolls be sold simultaneously. Independently, Benjamin Mays and Mordecai Johnson had reached the same conclusion. The key to the project's success lay in the dolls' ability to portray the diversity of black Americans.

Ideal's President David Rosenstein agreed. He had spent much of his life fighting poverty and prejudice. He tried to make his business an instrument of social change. He argued that the "toy world [was] indeed the whole world in miniature." The problem was that his partners were not social reformers. Rosenstein tried to convince them. He drew on the report of the Mid-Century White House Conference on Youth, where Kenneth Clark had presented his findings.[44] Michtom and Katz, however, did not think the Sara Lee Doll made good business sense, and they put the project "on the back burner."[45]

Rosenstein persisted. He asked Charlotte Klein, a public relations consultant, to help him. Klein remembers that during one of their periodic planning sessions she asked Ben Michtom which new products were coming up. Rosenstein seized the opportunity to make another pitch for the Sara Lee Doll. When Klein told Michtom and Katz that she thought the idea had commercial potential, they agreed to allow her to run a marketability study. She reported that "the Negro doll would make news for the company and outlined some publicity possibilities." It was decided that a few prototypes should be produced.[46]

Klein tested the doll herself. She boarded a New York subway train carrying the doll and noted passenger reactions. A young black woman approached her and asked, "Is this doll going to be sold? . . . It's the wrong color," the black woman declared. "It's too gray. That's the color of a *dead* baby."[47] Klein realized that Ideal needed expert advice in choosing the doll's color.

Ideal's delays frustrated Creech and von Hesse. Creech left the January 1951 meeting with Rosenstein convinced that the doll would soon be in production. Nine months passed and nothing had happened. In September Creech learned of Klein's impromptu market test and the color fiasco. She told Rosenstein that she wanted to come to New York, and Rosenstein arranged to pay her expenses for two weeks.[48] Creech arrived early in October. Two major obstacles blocked Ideal's production of the doll. Creech had to find a way to convince Michtom and Katz that the doll would make "news" for Ideal, and she had to find a solution to the color problem that Charlotte Klein had discovered on her subway ride. Creech and von Hesse had two weeks to come up with a solution.

Color Matters

Eleanor Roosevelt saved the Sara Lee Doll. On October 15, 1951, she invited Creech and von Hesse to visit her at the Park Sheraton Hotel in New York. Creech outlined what had transpired since their last meeting, including the support of black leaders and Klein's experiences on the subway train. Mrs. Roosevelt offered to hold a reception for the doll. Black and white leaders would be invited to a tea and would be asked to serve as a "color jury" for the doll. The reception would have to take place within ten days, since Mrs. Roosevelt was leaving for Paris shortly. The tea was scheduled for October 22.[49]

Mrs. Roosevelt asked the women to prepare a guest list by the next morning. It read like a who's who of African-American leaders. The final list included Charles Johnson, Mordecai Johnson, R. R. Wright, Jr., Helen Whiting, Lois Jones, Sadie Delaney, Benjamin Mays, Ralph Bunche, Walter White, Mary McLeod Bethune, James Hubert, Lester Granger, Jackie Robinson, Earl Brown, and Zora Neale Hurston.[50] They phoned David Rosenstein and told him about the reception. Rosenstein promised to mobilize his forces. He pledged that Ideal would have the models ready by Monday. Charlotte Klein went to work notifying the media.

Figure 11.1 Bernard Baruch, Eleanor Roosevelt, theatric producer John Golden, and Dr. James Hubert of the New York Urban League at Mrs. Roosevelt's "color jury," October 22, 1951. (COURTESY SARA CREECH.)

At five o'clock on October 22 Mrs. Roosevelt welcomed her guests and presented the "new colored doll, made by the Ideal Toy Corporation."[51] Roosevelt introduced Creech, von Hesse, and Rosenstein and outlined the "jury's" task. Members of the "color jury" were to examine the prototypes and select the color that they judged appropriate for the first Sara Lee Doll. "There were twelve to fifteen people in the room," Charlotte Klein remembers. "The dolls were displayed. There was lots of discussion. People would pass by each and discuss it in detail. I remember overhearing people say, 'Do you think this hits the mark.' They came up with a medium brown. No one wanted to make the doll too light. They fixed on a medium color."[52]

The reception persuaded Ideal's directors to authorize the doll's production. They knew that the press would give the Sara Lee Doll extensive coverage. The *St. Louis Post-Dispatch* ran a full-page story on the doll. *Time, Newsweek, Life Magazine, Ebony,* and *Independent Woman* also published stories.[53]

Initial sales were encouraging. Lothar Kiesow arranged for the doll to appear in Sears's 1951 Christmas catalogue. Large department stores in New York like Gimbel's and Abraham and Straus ran advertisements for the doll. David Rosenstein, who had just returned from Europe, was ecstatic. "It's high time," he declared in a press release "[that] there were [sic] a quality Negro doll which would give Negro children a new respect for their heritage and would give White children a new respect for the Negro."[54]

Rosenstein's and Creech's hopes for the doll's commercial success faded when high sales failed to materialize; however, they continued to believe in the doll as an educational tool. Miriam Gittelson, Rosenstein's secretary, succeeded in persuading the New York City public schools to purchase several hundred dolls annually for three years. The New York public school system was the single largest purchaser of the doll.[55]

There were several reasons for the doll's commercial failure. Color was central. First, there was a technical problem. In July 1951 Ideal introduced "Vinylite Magic Skin." Vinyl in the early 1950s was an unstable material. The secret to the Sara Lee Doll's "magic skin" lay in mixing plasticizers with synthetic resins. This gave the doll's skin softness. Katz and his chemical engineers discovered that over time the plasticizers seeped out of the vinyl. This caused the doll to harden, the skin to change color, and the doll's clothes to absorb the vinyl's dye.[56]

Second, Creech's desire to produce an anthropologically correct, quality black doll was impossible, as no one could define precisely what Zora Neale Hurston or anyone else meant by "anthropologically correct." Creech had set herself an impossible task. There is no such thing as an anthropologically correct black or white child. Years later Kenneth Clark identified this problem when he observed: "The serious research to correct the racial stereotyping and create the 'anthropological [sic] correct' doll is in itself an interesting concept. As if all black children look alike any more than all white children do. But the idea of providing a Negro child with other than a white doll was an advanced perception in the 1940s."[57]

What Creech, von Hesse, and Burlingame had intended to do was to create a doll that would depict black children with love and respect. This was a laudable objective. The questions raised about the doll's skin color, however, pointed to the fundamental contradictions involved in the project.

Ideal's officers ultimately sabotaged the Sara Lee Doll when they blocked production of Sara Lee's brother and sisters. Michtom and Katz let their commercial sense guide them. They did not share Rosenstein's and Creech's vision of toys as agents of social change. While the Ideal Company's interest in the doll peaked with Mrs. Roosevelt's tea, Creech considered the affair just a beginning. The day following the tea Creech wrote to the black educator Sadie Peterson Delaney, "Tomorrow I am working with them [Ideal] on additional models to follow this first one."[58] Creech also wrote to Benjamin Mays: "Today we had a conference with our manufacturer and he wants to have a little booklet printed to attach to each doll. It will contain a short history of the development of the doll and the purpose of the doll . . . to put into the hands of all children regardless of race a lovable Negro doll as fine and attractive as to quality and workmanship as any white doll."[59] Creech asked Delaney, Mays, Ralph Bunche, Eleanor Roosevelt, and others to contribute to the booklet.

Creech's relations with Ideal began to cool. She and von Hesse worried that Ideal's advertising director might misuse the materials they had prepared for the booklet. "Maybe we're over cautious," Creech declared, "but to be sure as possible that the material available to us is properly used, we are checking every bit of it . . . also, all of the publicity."[60]

Creech was right to be concerned. She and Michtom had already quarreled over Ideal's publicity campaign. Michtom had decided to use his own press agent. Creech was furious when the agent distributed an error-filled press statement. Michtom agreed that all releases would henceforth be cleared by Creech for accuracy. Three days later, however, Creech discovered that Michtom had authorized another batch of misleading press bulletins. She demanded a meeting with Michtom to set matters right, but she discovered she had nothing "to fall back on."[61] There was no contract with Ideal. After Mrs. Roosevelt's tea the doll had been rushed into production. Creech had no objections; she did not want Ideal to stop the forward momentum. She did, however, want an agreement that protected the doll's integrity. Creech hired a lawyer, Oliver Titcomb, to represent her interests, but he did not have much luck with Ideal's management.[62]

Late in 1952 a contract was signed, but by then Creech felt that Ideal had betrayed the project.[63] Commercialism had won out. Except for Rosenstein, Ideal's directors considered the doll a marketing gimmick. After the fanfare generated by Mrs. Roosevelt's tea, Ideal had no incentive to expand the project. Klein was instructed to curtail the publicity campaign, and the company abandoned the idea of including an educational booklet with the doll. In 1952 Ideal decided not to expand the Sara Lee line. Michtom and Katz favored other projects. In the comic strips, Ann Howe had married Joe Palooka. A *New Yorker Magazine* reporter asked Klein if anything was to be expected from Ideal. "Mr. Michtom will see to that," Klein responded.[64] The Joan Palooka Doll was born in 1953.

Ideal stopped producing the Sara Lee Doll in 1953. Sara Creech returned to Belle Glade to organize the first stage of a ten-year campaign to meet the day care and educational needs of migrant workers. Abe Kent, who was in charge of marketing at Ideal in the 1950s, recalled: "The summer of 1951 saw the introduction of the Bonny Braids doll. It was a fabulous success. Sara Lee didn't make it. I guess that tells you a great deal about how things go."[65] It would be another decade and a half before another toy manufacturer produced a quality African-American doll.

Despite the commercial failure of the Sara Lee Doll, Sara Creech's social consciousness and activism did not wane. She played a central role in the 1950s and 1960s in the struggle to provide day care for migrant workers. Her efforts led to the creation of the Wee Care Child Development Center in Belle Glade. In 1965 she testified before the White House Conference on Day Care. Although illness compelled her to leave Belle Glade in 1974 for the coastal climate of Lake Worth, her new home provided the opportunity to fulfill a dream of attending college. She enrolled in Palm Beach Community College and subsequently worked as a computer operator at the college until her retirement in 1985. Sara Creech remains active in efforts to forge interracial alliances.

NOTES

The author wishes to thank Sara Creech for permission to examine her private papers and Karen Lucas for her help in researching the history of the Sara Lee Doll.

1. Toni Morrison, *The Bluest Eye* (New York, 1970), 19–20.
2. Ibid., 22.
3. Interview with Sara Creech, May 10, 1993, notes in author's possession.
4. Mind over Matter," *Time,* November 5, 1951, 45.
5. Interview with Sara Creech, May 10, 1993.
6. "Modern Designs: Manufacturers Find Trends More Realistic," *Ebony* (January 1952), 46.
7. Interview with Sara Creech, May 10, 1993.
8. Ibid.
9. Maxeda von Hesse, "Notes from Mrs. Roosevelt's Tea," typescript, October 1952, Creech Papers, in possession of Sara Creech, Lake Worth, Florida.
10. Bruno Lasker, *Race Attitudes in Children* (New York, 1924), 220.
11. Ibid., 220–21.
12. W. D. Weatherford, *The Negro from Africa to America* (New York, 1924), 427.
13. Lasker, *Race Attitudes in Children,* 220.
14. "Negro Dolls Popular with the Public Since Birth in 1919," *Ebony* (January 1952), 49.
15. Ibid.
16. "Black Dolls: 1840–1990," 4, typescript for exhibit at Wenham Museum, Wenham, Massachusetts, February 1–April 7, 1991.
17. Lasker, *Race Attitudes in Children,* 221.
18. "Negro Dolls Popular with the Public," 49.
19. Hesse, "Notes from Mrs. Roosevelt's Tea."
20. Sara Creech, Maxeda von Hesse, and Sheila Burlingame, "The Story of How the Sara Lee Doll Came to Be," 3, typescript, Creech Papers.
21. Ibid., 2–3.
22. Interview with Sara Creech, May 10, 1993.
23. Interview with Sara Creech, October 8, 1993, notes in author's possession.
24. Ibid. This expression comes from Hurston's experiences collecting folklore in railroad camps. When locomotives were put into reverse, a bell rang to warn workers. Hurston used the phrase to

mean, "Don't retreat, don't look back, go ahead with your business." See Interview with Sara Creech, December 4, 1993, notes in author's possession.

25. Sara Creech to Benjamin Mays, October 23, 1951, Creech Papers.

26. Creech, et al., "Story of How the Sara Lee Doll Came to Be," 5.

27. Telephone interview with Sara Creech, June 8, 1994, notes in author's possession.

28. Zora Neale Hurston to Creech, June 29, 1950, Creech Papers.

29. Mays to Creech, November 9, 1950, Creech Papers.

30. Hurston to Creech, June 29, 1950.

31. Thirty years earlier Hurston had spent many evenings at Georgia Douglas Johnson's halfway house taking part in "marathon literary discussions." See Robert E. Hemenway, *Zora Neale Hurston: A Literary Biography* (Urbana, 1977), 19.

32. Creech, et al., "Story of How the Sara Lee Doll Came to Be," 5.

33. Eleanor Roosevelt to Creech, August 8, 1850, Creech Papers.

34. Creech, et al., "Story of How the Sara Lee Dolls Came to Be," 5.

35. Ibid., 6.

36. Telephone interview with Miriam Gittelson (personal secretary to David Rosenstein), November 1, 1993, notes in author's possession.

37. Interview with Sara Creech, October 23, 1993.

38. Mordecai Johnson to David Rosenstein, February 21, 1951, Creech Papers.

39. Kenneth Clark to author, May 26, 1994, in author's possession.

40. Richard Kluger, *Simple Justice* (New York, 1976), 318.

41. Charles Johnson to Creech, February 6, 1951, Creech Papers.

42. Beryl Shelton, "The Sara Lee Dolls: Summary of Spontaneous Impressions of 20 Children and 10 Adults at Fisk University Nursery School," 6, typescript, Creech Papers.

43. Johnson to Creech, February 6, 1951.

44. Kenneth Clark, *Prejudice and Your Child* (Boston, 1955). Clark published an expanded version of the material that he presented in the White House conference in this book.

45. Telephone interview with Charlotte Klein, October 13, 1993, notes in author's possession.

46. Charlotte Klein to author, October 25, 1993, in author's possession.

47. Telephone interview with Charlotte Klein, October 13, 1993.
48. Creech to Harley Kimmel, November 5, 1951, Creech Papers.
49. Interview with Sara Creech, May 10, 1993.
50. "Guest List for Mrs. Roosevelt's Tea," typescript, Creech Papers.
51. Watertown, NY, *Times,* October 24, 1951.
52. Telephone interview with Charlotte Klein, October 13, 1993.
53. Ibid.; *St. Louis Post-Dispatch,* November 11, 1951; "Mind Over Matter," 5; "In Passing," *Newsweek,* November 5, 1951, 48; "Doll for Negro Children: New Toy Which is Anthropologically Correct Fills an Old Need," *Life,* December 17, 1951, 61–62; "Modern Designs," 46–48; Gertrude Penrose, "Mission of a Doll," *Independent Women* (December 1951), 350–51.
54. "Realistic Negro Dolls to Combat Racial Prejudice in Youngsters," typescript of press release from Edward Gottlieb & Associates, n.d., Creech Papers.
55. Telephone interview with Miriam Gitttelson, November 1, 1993.
56. Judith Izen, *Collector's Guide to Ideal Dolls* (Paducah, KY, 1994), 8.
57. Kenneth Clark to author, June 16, 1994, in author's possession.
58. Creech to Sadie Peterson Delaney, October 23, 1951, Creech Papers.
59. Creech to Mays, October 23, 1951.
60. Creech to Harley Kimmel, November 5, 1951.
61. Ibid.
62. Oliver S. Titcomb to Creech, March 27, 1952, Creech Papers.
63. Interview with Sara Creech, May 10, 1993.
64. Izen, *Collector's Guide to Ideal Dolls,* 22.
65. Telephone interview with Abe Kent, November 1, 1993, notes in author's possession.

AFTERWORD: RETHINKING TECHNOHISTORY
Chris Hables Gray

TECHNOLOGY IN HISTORY

From the Sugar River in New Hampshire or a small doll in Florida through colorizing *Casablanca* to the informatics of nursing and the planning of future space explorations might seem a fair distance for some, but it is the necessary terrain of the historian and, as we have seen in this collection, of the writing teacher, the philosopher, the social psychologist, the professor of nursing, and the communications theorist as well. Only unfortunate circumstances prevented the participation of several other contributors from still other disciplines (engineering, Spanish, and journalism) who attended the 1992 NEH summer seminar in Cleveland on "Technology and American Culture" and who wished to write for this collection. Note also that I assume that the history of technology is part of every historian's domain.

Why is it necessary that every historian take the history of technology into account? Because it is Man the maker, the toolmaker (human the toolmaker, really). *Homo faber.* It is not just that without technology civilization would be impossible, it is more. Without technologies (tools with techniques), humans—even living the most pristine hunter-gather lifestyle—cannot survive. The great historian of technology, Lynn White, Jr., argued a similar point, although with a bit more restraint.

> The new school of physical anthropologists who maintain that *Homo* is *sapiens* because he is *faber,* that his biological differentiation from the other primates is best understood in relation to tool making, are doubtless exaggerating a provocative thesis. *Homo* is also *ludens, orans,* and much else. But if technology is defined as the systematic modification of the physical environment for human ends, it follows that a more exact understanding of technological innovation is essential for our self-knowledge.

Once agriculture was invented and then cities, the symbiosis between life and (techno) culture (culture relies on technology, of course) became so intimate that it is usually ignored, especially by historians. Then came machines, perhaps first made up only of people and weapons organized into armies, as Lewis Mumford speculated, or perhaps the simple levers and pumps of agricultural works had primacy, we don't really know. In any event, machines, and metaphors of machines, spread quickly so that within a few thousand years many would speak confidently of Man the Machine, the Clockwork Universe, and finally the Machine Age.

Recently we have seen rapid shifts in the machine metaphor, reflecting different generations of machines. From the clock of the Middle Ages to the power-producing plumbing labyrinth known as the steam engine that loomed over the Industrial Revolution to the information processing machine, the computer, that reigns today, machines have changed and spread. And, in incredible ways, the relationship between humans and our tools/machines has changed as well. Now it is no longer *homo faber* but rather *faber homo faber,* manufactured man making man, or better manufactured humanity making humanity, or perhaps *homo cyborg,* or even *cyborg faber,* cyborg the maker.

This is especially true of American culture. Many have called America "technology's nation," as Thomas P. Hughes has pointed out. Well, nothing could be more American than a cyborg. Mark Seltzer remarks in his article on American male bodies, "The Love Master":

> It might be argued that nothing typifies the American sense of identity more than the love of nature (nature's nation) except perhaps its love of technology (made in America). It's this double discourse of the natural and the technological that . . . makes up the American "Body-Machine complex."

So, perhaps we now live in the Cyborg Age. Notice that with the passing of the Stone Age, the Iron Age, the Machine Age, and the Industrial Age, stones, iron, machines, and industry did not really go away. Rather, new phenomena were layered over them. I don't see how cyborgism can now go away, barring apocalypse . . . but in the future, no doubt, it will be less noticeable, since it is now that cyborg relations are exploding into culture; exploding because of the many sites implicated, the great speeds involved, and the cultural dislocation that is inevitable with the emergence of the cyborg society.

So it is with technology in general. Old technologies are so ubiquitous that they are transparent. Building, writing, personal grooming, animal husbandry, weapons, and other ancient technologies are often treated by historians as if they were as natural as the sky and the oceans, instead of being integral and created parts of human culture, and therefore our human history.

Technology is clearly a fundamental part of being human, and yet in many respects the history of technology is ghettoized. In fact, the Society for the History of Technology (SHOT), a surprisingly friendly and productive academic association, was formed originally because the historians who controlled the History of Science Society didn't want issues of technology muddying up the crystalline streams of pure science where they fished for historical insight.

Now the history of science is certainly important, although science is really quite a recent and specific part of the history of ideas and, in part, of the history of technology. Indeed, one could easily argue that science's greatest importance is based on what it has done to technology, for as a methodology for producing technologies, science is unexcelled. Therefore we can thank science for a number of things including television, vaccinations, hospitals, postmodern war, and space travel. But we do well to remember that technology predates and encompasses science, and yet it is considered a historical field of secondary interest and second-class status.

Why is this so? Is it the arrogance of mind and abstract learning over craft knowledge? Yes, in part, that is the short answer, but the real problem is disciplinarity. The real question here is, Why is academia set up as if knowledge falls into discrete (yet obviously highly arbitrary) categories?

DISCIPLINARITY

The academic disciplines are rather a new phenomena. Until the last hundred years or so, if you were interested in ethics, or the principles of politics, or what defined animals, or history, you were a philosopher. Perhaps there was a distinction drawn between "natural" philosophers who were more interested in nature, from other philosophers who looked most closely at humans and their culture, but there were no borders between inquiries, except around God and his theologies. With the rise of research universities in Germany and then the United States, this changed. Not only were specific disciplines delineated but each one was given a particular status, starting with physics in the natural sciences and ending with the history of technology in the humanities, or so it sometimes seems.

Certainly the "hard" sciences have more status than the "soft" social sciences and they have more than the very soft humanities and the hierarchy ends with the arts, which many scientists accuse of lacking any discipline whatsoever. But the term *discipline* means more than hard focused work actually, for many an artist works harder than many a scientist. What is actually being "disciplined" into discrete, manageable realms is knowledge itself. Yet it seems that knowledge continually struggles against captivity. There is no consensus on what constitutes a discipline. Is it a field? An object of study? A problem area? A methodology? A set of colleagues? An academic association? A pattern of citations? And why are there so many of them? According to Warren Hagstrom, in 1918 there were 149 fields in the United States in which graduate degrees were awarded; by 1960 there were 550. No doubt there are even more today.

Perhaps the main reason for the proliferation of disciplines is that knowledge refuses to lie in any Procrustean bed made by academics. Certainly, that is the reason that work that goes beyond the disciplines has proliferated. Call it what you will, transdisciplinary, multidisciplinary, interdisciplinary, or whatever, many scholars see their work as transcending the typical boundaries of their home discipline, and some even admit it. One of the goals of this book has been to demonstrate that good academic work is *by necessity* not limited to any one discipline.

This isn't the place to go into the history of interdisciplinary research which includes the creation of "area" studies earlier in this century and of "ethnic" and "gender" studies more recently. I won't even go into the proliferation of science and technology studies (STS) where many historians of technology have found a home. Julie Thompson Klein, for one, has written a fine account of all of this in her book *Interdisciplinarity: History, Theory, and Practice,* where she also explains the differences between trans-, inter-, and multidisciplinarity and of "integrative" studies.

What is important for the argument here is the impetus behind challenges to confining assumptions about disciplined knowledge. Often, it is assumed that interdisciplinary research is harder than staying within one discipline (which is true) and that only a few scholars should attempt it (which is quite untrue). Yes, comparative studies involves more work than just looking at one culture or type of literature. Yes, biochemistry means knowing a great deal about

Afterword

biology and chemistry, not just the one or the other. Yes, a good historian of American culture has to know about many different technologies and their uses. But that is life. The living world is very complicated and to understand it involves much work and much complex thinking. It means being willing to go beyond one's simple discipline and incorporating the hard-won insights of other fields.

But this doesn't mean that a good scholar has to excel in every field, something that is quite impossible today. To be honest, the contributors to this book have not all excelled in the writing of the history of technology, for example. Nor have we all discovered and used new sources or new analytical frameworks with increased explanatory (or is it seductive?) power. Some of us have, actually, in my opinion, but no, most of us didn't write brilliant history of technology, but we used good history of technology to try and do other things, broader things: to write local history, to teach English, to explain the thinking and feeling behind cutting-edge virtual reality research, to reveal something about how our mass media is controlled, to help formulate public policy, and so on. Because history, or any work in academia, should not be done just to do it, whatever it is, better. It is to understand the world better, and in a small way I'm sure, to make the world better.

Julie Klein remarks in the conclusion of her book on interdisciplinarity that:

> Certain character trait have ... been associated with interdisciplinary individuals, among them reliability, flexibility, patience, resilience, sensitivity to others, risk-taking, a thick skin, and a preference for diversity and new social roles. The tendency to follow problems across disciplinary boundaries is, in fact, seen as a normal characteristic ...

I would submit that these are admirable qualities for *all* scholars to have, not just those involved in self-consciously interdisciplinary work. As my father's father, a mechanic, used to say, if it is worth doing, it is worth doing "all the way through."

TECHNOHISTORY: TECHNOLOGY AND HISTORY

Why use the term *technohistory,* which surely must annoy some academics, especially those who view themselves as somewhat more traditional? It goes back to the point made (no doubt crudely and repetitively) at the start of this essay—technology is *integral* now, not just to history, but to humanity. That is why today there is a proliferation of "techno" books and concepts. There is "technoculture" which describes contemporary America, there is "technopolis" to label our technology-dependent cities, there is "technobabble" to talk about today's discourse, jammed with technological jargon. The value of the term *technohistory* is that it makes the centrality of technology clear, even unavoidable. It isn't really just the history of technology, it is history in the context of technology, it is *Technology & Culture* as the Society of the History of Technology's journal is entitled. And, as Carroll Pursell argues in his Introduction, "Reclaiming Technology for the Humanities," it is a viewpoint the technologists need to incorporate as well.

Our work hopes to defend a broader understanding of just what the history of technology is, harkening back as it does to the original emphasis of SHOT, as is made manifest in its journal's name. Understanding the relationship between technology and human culture is both a crucial academic question and one of great social importance. Through these case studies we have endeavored to show that it is not a simple question answered by any one discipline or through any one particular historical approach.

We are open to the insights of new ways of looking at history. Notice I didn't say we align, or side ourselves, with the new history (often labeled "postmodern"). All of us who contributed to this book, and all of us at the seminar in fact, feel a great sympathy, maybe even reverence, for history itself—the idea of trying to understand humanity and our works by what our ancestors have already done. And we honor the work that traditional historical approaches (and there are actually many) have accomplished. But we are open to the new insights from feminists, poststructuralists, and deconstructionalists as well. What we hope to accomplish is to integrate some of these new and different ways of understanding our collective past with the more traditional historiographical approaches.

Perhaps this open-mindedness was only possible because our host, Professor Pursell, is very broad-minded himself and because most of the seminar participants were not "officially" historians and therefore were free of preconceived historiographical positions? Perhaps it was just Carroll Pursell's nose for tolerant and open-minded academics with an impulse toward synthesis instead of toward intellectual triumph? In any event, we found in our mutual conversations a great deal to agree about, and we even found pleasure, and insight, in our disagreements.

John Staudenmaier, in his history of SHOT, notes that from the

Afterword

beginning historians of technology have been arguing about methodology. To this day there is still no consensus, which to my mind is how it should be because there is no one golden way to understand history. But there has been progress. As Staudenmaier shows, detailed "internalist" accounts of the workings of complex machines have been supplemented over the years by "externalist" analysis of the cultural situation of those artifacts and by the "contextualist" merging of both internalist technical detail and externalist cultural understandings. Most recently, SHOT has consciously sought to expand on its analysis of American, and even Western technology, to explore technohistory's worldwide implications. Staudenmaier points this out with admirable precision:

> In our present circumstances—I refer to the late twentieth century and to those cultures generally called 'the West'—the profound influence of technology poses a challenge to the historical endeavor. It is a commonplace observation that increasingly complex technological networks have dramatically changed the cultural world of the West. Equally important, the extraordinary westernization of technologies in other societies has created a cultural problem involving the entire planet. For both reasons the contribution of the historian of technology is critical. How will we, in the West, tell the tale of our technological past? What interpretative language will shape our frame of reference for thinking about and responding to current technological issues? How will we understand the relationship between Western technology and nonwestern societies?

How indeed? Perhaps the greatest drawback of this collection is that it hardly begins to explore this relationship between Western technology and non-Western societies. Comparative work is more difficult (in all sorts of ways) than just staying home and studying what one finds in the backyard. So it is that among the articles that didn't get written for this collection were most of those that look at technology in a comparative context. Distance and domestic disturbances kept one seminar participant from writing on how American ideas of appropriate technology apply, and don't apply, to his homeland of Ethiopia. Difficulties in time and travel kept articles on Western communications technology in Taiwan and on Panamanian control of the canal from being written. A lack of accessible resources prevented an article on black inventors and mechanics in the Southern states of the eighteenth and nineteenth centuries from being produced on time and it was the same with a comparison of the role of bicycles in the United States, China, and Cuba. But this is to be expected. Maybe in the future these studies will get completed and published. At least we tried.

It is good to set high goals, but it is crucial to have obtainable goals as well. In his letter explaining the seminar to potential participants, Carroll Pursell wrote,

> It seems to me that it is desirable that the literature and methodology of the history of technology be shared more widely both with historians and with scholars in other academic fields as well.

If one adds "students" and "the general public" to Carroll's list, that would make a pretty fair statement of the primary goal of this collection. I hope we have met it.

BIBLIOGRAPHICAL NOTE

The quotations in this essay can be found in the following texts, alphabetized by author. All emphasis in the quotations is the same as in the originals.

Warren Hagstrom, *The Scientific Community* (New York: Basic Books, 1965), p. 220; Thomas P. Hughes, *American Genesis: A Century of Invention and Technological Enthusiasm* (New York: Penguin Books, 1989), p. 2; Julie Thompson Klein, *Interdisciplinarity: History, Theory, and Practice* (Detroit: Wayne State University Press, 1990), p. 183; Mark Seltzer "The Love Master" in Joseph A. Boone and Michael Cadden, eds., *Engendering Men: The Question of Male Feminist Criticism* (New York: Routledge, 1990), pp. 141–2; John M. Staudenmaier, *Technology's Storytellers: Reweaving the Human Fabric* (Cambridge, MA: MIT Press & SHOT, 1989), p. xiii; Lynn White, Jr., "The act of invention: Causes, contexts, continuities, and consequences," *Technology & Culture,* vol. 3, no. 4, Fall 1962, p. 498.

SUGGESTED READINGS IN TECHNOHISTORY

Ahmed, I., ed. (1985) *Technology and Rural Women.* London: Allen and Unwin.
Beniger, J. (1986) *The Control Revolution: Technological and Economic Origins of the Information Society.* Cambridge, MA: Harvard University Press.
Bijker, W., Thomas Hughes, and Trevor Pinch, eds. (1987) *The Social Construction of Technological Systems: New Directions in the Sociology and History of Technology.* Cambridge, MA: MIT Press.
Boyer, Paul (1985) *By the Bomb's Early Light: American Thought and Culture at the Dawn of the Atomic Age.* Pantheon: New York.
Cockburn, Cynthia (1983) *Brothers: Male Dominance and Technological Change.* London: Pluto Press.
Commoner, Barry (1971) *The Closing Circle: Nature, Man, and Technology.* New York: Knopf.
Cooley, M. (1980) *Architect or Bee? The Human/Technology Relationship.* Slough, England: Langley Technical Services.
Corn, Joseph (1983) *The Winged Gospel: America's Romance with Aviation, 1900–1950.* Oxford University Press: New York.
Corn, Joseph, ed. (1986) *Imagining Tomorrow: History, Technology and the American Future.* MIT Press: Cambridge, MA.
Cowan, Henry J. (1977, 1985) *The Master Builders: A History of Structural and Environmental Design from Ancient Egypt to the Nineteenth Century.* Melbourne, FL: Krieger Publishing.
Cowen, R. S. (1983) *More Work for Mother: The Ironies of Household Technology from the Open Hearth to the Microwave.* New York: Basic Books.
Cutcliffe, Stephan, and Robert Post, eds. (1988) *In Context: History and the History of Technology—Essays in Honor of Melvin Kranzberg.* Bethlehem, PA: Lehigh University Press.
Dickson, David (1974) *Alternative Technology and the Politics of Technical Change.* Scotland: Fontana.

Ellul, Jacques (1964) *The Technological Society,* trans. by J. Wilkinson. New York: Knopf.
———. (1980) *The Technological System,* trans. by J. Neugroschel. New York: Continuum.
Eubank, Keith (1991) *The Bomb.* Melbourne, FL: Krieger Publishing.
Faulkner, W., and E. Arnold, eds., (1985) *Smothered by Invention: Technology In Women's Lives.* London: Pluto Press.
Hacker, S. (1989) *Pleasure, Power and Technology.* Boston, MA: Unwin Hyman.
Hindle, Brooke, ed. (1975) *America's Wooden Age: Aspects of its Early Technology.* Sleepy Hollow Restorations: Tarrytown, NY.
Hindle, Brooke, and Steven Lubar (1986) *Engines of Change: The American Industrial Revolution, 1790–1860.* Washington, DC: Smithsonian Institution Press.
Howe, Barbara J., and Emory Kemp, eds. (1986) *Public History: An Introduction.* Melbourne, FL: Krieger Publishing.
Hughes, Thomas (1983) *Networks of Power: Electrification in Western Society, 1980–1930.* Johns Hopkins University Press: Baltimore.
———. (1989) *American Genesis: A Century of Technological Enthusiasm, 1870–1970.* New York: Penguin Books.
Ihde, Don (1990) *Technology and the Lifeworld: From Garden to Earth.* Bloomington, IN: University of Indiana Press.
Karmarae, Cheris, ed. (1988) *Technology and Women's Voices.* New York: Routledge & Kegan Paul.
Kasson, John (1976) *Civilizing the Machine: Technology and Republican Values in America.* Grossman: New York.
Miller, Ron (1993) *The Dream Machines: An Illustrated History of the Spaceship in Art, Science and Literature.* Melbourne, FL: Krieger Publishing.
Mitcham, Carl, and Philip Siekevitz, eds. (1989) *Ethical Issues Associated with Scientific and Technological Research for the Military.* New York: New York Academy of Sciences.
Miles, I. (1988) *Home Informatics: Information Technology and the Transformation of Everyday Life.* London: Pinter Publishers.
Morison, Elting E. (1974) *From Know-how to Nowhere: The Development of American Technology.* Basic Books: New York.
Mumford, Lewis (1934, 1963) *Technics and Civilization.* New York: Harcourt, Brace, & World, Inc.
———. (1966) *The Myth of the Machine Volume One: Technics and Human Development.* New York: Harcourt, Brace, & World, Inc.

———. (1970) *The Myth of the Machine Volume Two: The Pentagon of Power.* New York: Harcourt, Brace, & World, Inc.
Noble, David (1977) *America by Design: Science, Technology, and the Rise of Corporate Capitalism.* Knopf: New York.
———. (1984) *Forces of Production: A Social History of Industrial Automation.* New York: Knopf.
O'Connor, John E., ed. (1990) *Image as Artifact: The Historical Analysis of Film and Television.* Melbourne, FL: Krieger Publishing.
Pacey, A. (1983) *The Culture of Technology.* Oxford: Basil Blackwell.
Pursell, Carroll W., ed. (1991) *Technology in America: A History of Individuals and Ideas.* Cambridge, MA: MIT Press.
Reiser, S. (1978) *Medicine and the Reign of Technology.* Cambridge, England: Cambridge University Press.
Rothschild, J., ed. (1983) *Machina Ex Dea: Feminist Perspectives on Technology.* New York: Pergamon Press.
Sinclair, Bruce, ed. (1986) *New Perspectives on Technology and American Culture.* Philadelphia: The American Philosophical Society.
Smith, Merritt Roe, ed. (1985) *Military Enterprise and Technological Change; Perspectives on the American Experience.* MIT Press: Cambridge, MA.
Smith, Merritt Roe, and Leo Marx, eds. (1994) *Does Technology Drive History?* Cambridge, MA: MIT Press.
Solomonides, T., and Les Levidow, eds. (1985) *Compulsive Technology: Computers as Culture.* London: Free Association Books.
Stanworth, M., ed. (1987) *Reproductive Technologies; Gender, Motherhood and Medicine.* Cambridge, MA: Polity Press.
Staudenmaier, John (1985) *Technology's Storytellers: Reweaving the Human Fabric.* Cambridge, MA: MIT Press.
Stover, Carl, ed. (1963) *The Technological Order.* Detroit: Wayne State University Press.
Trescott, M. M., ed. (1979) *Dynamos and Virgins Revisited: Women and Technological Change in History.* Metuchen, New Jersey: Scarecrow Press.
Turkle, Sherry (1984) *The Second Self: Computers and the Human Spirit.* New York: Simon & Schuster.
Van Creveld, Martin (1989) *Technology and War: From 2000 B.C. to the Present.* New York: Free Press.
Wajcman, Judy (1991) *Feminism Confronts Technology.* University Park, PA: Pennsylvania State University Press.

Winner, Langdon (1986) *The Whale and the Reactor: A Search for Limits in an Age of High Technology.* Chicago: University of Chicago Press.

Wright, Barbara Drygulski, et al., eds. (1987) *Women, Work, and Technology: Transformations.* Ann Arbor: University of Michigan Press.

Yoxen, E. (1986) *The Gene Business: Who Should Control Biotechnology?* London: Free Association Books.

Zimmerman, J., ed. (1983) *The Technological Woman: Interfacing with Tomorrow.* New York: Praeger.

———. (1986) *Once Upon the Future: A Woman's Guide to Tomorrow's Technology.* New York: Pandora.

INDEX

All texts and bibliographies are indexed. *Italicized* numbers indicate it is an illustration or figure. *Italicized* words indicate a foreign language, a book, a periodical, a film, or a work of art. Note citations are given in the format 1n (note on page 1).

Academy of Motion Picture Arts and Sciences, 23n
Accidents, 45
  nuclear, 36
  potential, 47
  radioactive waste, 53
ACM, *see* Association of Computer Manufacturers
Acoustic telegraphy, 128
Actors, electrodigitalized, 186
AEC, *see* Atomic Energy Commission
African-American
  children, 233–49
  leaders and the Sara Lee Doll, 238
Ages (stone, iron, machine, cyborg), 252
  cyborg relations, 154
Allen, Woody, 16, 18, 24
  colorization, 7, 12
Allied Grand Doll Manufacturing Company, 236
Altman, Laurence K., 148, 149, 169n, 170n, 171, 172
Alwall, Dr. Nils, 147, 166, 172
America(n)
  allegiance to private property, 7
  character, 134
  colonial settlement, 97
  cosmetic surgery, 142
  culture
    and cyborgs, 252
    historians of, 255
  fear of surprise attack, 221–2
  frontier metaphor, 217; decline of, 222–4
  language, unique, telegraphic, 123
  self-reliance of, 217
  sense of discovery, 216
  society, postmodern and multicultural, 228
  space flight, acceptance, 218; vision of, 220
  steam engine, first, 101; spread of, 102
  technologies, new, 82
American Hospital Supply Corporation, Symbion stockholder, 170n
Americans for Rational Energy Alternatives (AREA), 50
American Society for Artificial Internal Organs (ASAIO), 153
American Society of Cinematographers (ASC), 13n
American Telephone and Telegraph Co., *see* AT&T
Amputee(s), 145
  in ancient Greece, 144
  and artificial limbs, 157
  medical, 165
  and prosthetic improvements, 163
Analog
  coloring system, 10
  computing, 9

Analogy as a tool, 216
Animal
    baboon liver transplants, 170n; as organ donors, 169n
    calves, artificial hearts, 171n
    chimpanzee, transplants, 169n
    data, falsified, 171n
    dogs, artificial hearts, 163
    experimentation, 149, 169n
    experiments with nuclear hearts, 170n
Annihilation, of space through time, 182
Anthropology, cyborg, 142
Apollo Project, 222
AREA, see Americans for Rational Energy Alternatives
Area studies, 254
Art
    in a battle with the government, 15
    European conceptions of, 7
    film is, Congress says, 18
    high, assumed authority, 22; popular transformed as, 22
    of impression management, 134
    and money, 17
    nursing is an, 72, 73
    popular transformed to high, 22
    works, 21
Artifacts, politics of, 212
Artificial
    blood, 143
    cells, 143
    cochlea, 143
    genitalia, 144–6, 158, 169n
    hands, 165; early, 171n
    hearts, 150–4, 170n
        bridges, 153, 171n
        economics of, 151–2
        ethics violations, 152
        plutonium power, 170n
        politics of, 152–3
        psychological costs, 152–3
        research on calves, 149, 171n
        Symbion, 171n
        totally implantable (TIAH), 150
    implants, 141, 143
    intelligence, 166, 192n
    joints, 143
    kidneys, 147–8; history of, 166
    limbs, culture of, 157; history of, 144–6
    livers, 147, 149
    organs
        commitment to, 159
        geographic distribution, 166
        hybrid, 148–50
        and natural organs, 164
        and science fiction, 162
    penis, 158, 169n
        Pearman's, 146
        Scott's inflatable, 146
Artistic copyright, 16
ASAIO, see American Society for Artificial Internal Organs
ASC, see American Society of Cinematographers
Association of Computer Manufacturers (ACM), 145, 171
Association of Independent Television Stations (INTV), 13
Association of Motion Picture and Television Producers, 24n
Atomic Energy Commission (AEC)
    dismantled, 36
    nuclear power for artificial hearts, 170n
    scandals, 36
Atoms for Peace, 35
AT&T (American Telephone and Telegraph Co.)
    ads, 90
    and democracy, 92
    grown, 90
    invents loading coil, 128, 138n
    and The Philistine, 89–93
    and symbols, Egyptian, 93
    telephones, 94
    versus Western Union, 129
    women as stockholders, 90

Index

Automation
  and cyborgs, 211
  and nursing, 75
  in World War II, 129
Autonomy, cyborg, 211

Baboon
  transplants, 169n
  hearts, 148–9
  liver, 149, 170n
Baby Faye Case, 148–9
Bell, Alexander Graham, 127
Bell Labs, 138n
Bell System, 120. For telephone, see AT&T
Bingaman, Senator Jeff, 43, 45
Bioethics, 154
Biofeedback, see Feedback, bio
Biomaterials, 143
Biomedicine, see Medical/Medicine, bio
Blacks, see African-American
Body/Bodies
  articulated through technology, 179
  centrality, 180
  cyborgian, 208; not innocent, 168
  displacement, 185
  human as machine, 167
  hybrid, 208
  image
    attitudes toward technology, 158
    modification of, 156
    new, 184
  integrity, adequacy of, 157
  modification, 156
  organic, 208
  perfection, 144
  pleasures, 184
  reinvention, 188
  remaking, 142
  resisted, 184
  substrate, 185
  technique of, 179
  technologized, 182
Bunche, Ralph, 234, 240, 243

Bureau of Land Management, 45
Bush, President George, 215
  and WIPP, 46
Business
  big, and *The Philistine,* 93
  first based on electronic communication, 122
  science, 84–5
  wealth, 84–5
Business and Professional Women's Clubs (BPWC), 234

CANT, see Citizens Against Nuclear Threat
Capital, as power/knowledge, 181; corporate, 121
Capra, Frank
  against colorization, 7
  colorization deal 11, 12
  heart, 14
CARD, see Citizens Against Radioactive Dumping
Carlsbad, New Mexico, 33
  Environmental Monitoring and Research Center, 57
  frustrations, 57
  mayor of, 56; testimony of, 64n
  and potash industry, 55
  salt beds, 55; mining of, 37
  and WIPP, 55–7; testing criticized, 56
  see also Radioactive waste; Waste Isolation Pilot Project
Carter, President Jimmy
  administration, 37, 39, 41
  Interagency Review Group, 38
Casablanca, 8, 28; colorization, 6; didn't need, 7
CCNS, see Concerned Citizens for Nuclear Safety
Cinema, see Motion pictures
Citizens Against Nuclear Threat (CANT), 49, 50
Citizens Against Radioactive Dumping (CARD), 43
Citizens for Alternatives, 49

Clark, Barney, 152
  artificial heart; failure, 170n; feeling of, 167
  experience 169n
  motivation, 170–1n
Clark, Kenneth, 242, 248n; and Sara Lee Doll, 245
Class
  futurism, 203
  and language, 137n
  and medical care, 166
  in nursing, 66–7
  and technology, 209
Coalition for Direct Action at WIPP, 49
Cold War, *see* War, Cold
Color(ization)
  aesthetic attacks on, 13
  alliance of studios and labs, 14
  bottom line, 12
  business factors, 14, 23n
  as catalyst, 5–7
  commercializing, 7–15
  digital; future, 23; invented, 11
  of dolls, 238
  economics of, 12, 13, 23n
  history of, 9
  importance, 5
  invented, 7
  is about movies, only movies, 22
  is about television, 12
  as localized dispute, 7
  market, 23n; created, 14; alliances, 17
  of skins, dolls and people, 233–49
  videotapes, 12; copyrighted, number of, 31–2
Colorization, Inc., 9, 14
  inactive, 23n
Color Systems Technology (CST), 10, 14, 31
  Entertainment Imaging, Inc, 23n
Committee to Make WIPP Safe, 52
Communication
  abstraction, 128, 132–5; mathematical, 127–9

  accuracy, 130
  for control, 121–2; precision, 130
  electrical effect, 119
  engineering, 135
  and language, 119–39
  mathematical, 127–9; physics, 129
  process, 130
  and self, 119–39
Communist Manifesto
  first English translation, 199
  gothic ghosts, 199
  and women, 212n
Communist manifestos, 196–202
Community, ideal, 98
Computer(ization)
  analog, and colorization, 9
  animation, 11, 23
  and change, 69
  cyberspace vehicle, 183
  in hospitals, 69
  and human love affair with them, 189
  and human nervous system, 130
  and nursing, 74
  and scientists, 131
  systems, 187; overpromised, 75
Concerned Citizens for Nuclear Safety (CCNS), 51, 54
Congress, U.S.
  acknowledges film is art, 18
  elite, 15
  librarian, 18; report, 19
  radioactive waste funding, 38; compromise, 45; WIPP, 42
  WIPP, authorization, 42
  see also Library of Congress
Contextualist history of technology, 257
Control
  through communication, ix, 119, 121–2
  crisis of, 121
  miniaturized systems, 151
  theory, 120

Index

Cooley, Dr. Denton, 150, 152
  ethical lapses, 153, 171
Courts, see U.S. District Court; Washington District Court
Creech, Sara Lee, 233–49
  against prejudice, 234
  dolls, history of, 235–7
  and Ideal Toy Co.; cooling relations, 245; delays with doll, 242–3
  and Inter-Racial Council, 235
  later career, 246
  migrant worker campaign, 246
  papers, 247
  Sara Lee Doll; Ideal Toy Co., 242–5; origins, 233–4
Cybercitizens, 184–5
Cybernauts, 190
Cybernetics, 129–32, 142
  biomedicine, , 136n
  feedback, 168n
  foundations, 129
  origins, 120; see also Information science
  robotics, 136n
Cyberobjects, 186
Cyberspace, 179–80
  defined, 180–3
  discourses, 184
  embodiments, 185–9
  erotic ontology, 189
  as habitat, 188–9
  meaning, 190–1
  mechanisms, 186
  metaphors, 180, 183–5, 187
  military technology, 181, 183
  as new space, 189
  Oz-like, 183
  paradoxical structures, 187
  psychology of, 185–6
  technology, 179–80, 183
Cyborg, 130
  age, 252
  American, 252
  anthropology, 142
  becoming, 188

body, not innocent, 168
defined, 132, 136, 168n, 213n; definitions, 208; meaning, 211
desire for, 159–66
different from humans, 166–7
downloaded, 166
economic reasons for, 160, 162
enhanced, 165; killer, 168
envy, 188
ethics, 163; good and bad, 168; rich vs. poor, 166
exhibitionism, 158
faber, 252
figure, 209; image, 187
future, 164–6; immortality, 167
genealogy, 143; history of, 165; origins of word, 168n
geographical distribution, 141; proliferation, 141, 148
healing, 161
importance of, 142; justifications for, 160–1; personal, 161–4
-ism, 142
manifestos, 209
material, 207
medicine, ix, 144
  forces behind proliferation, 159
  justifications, 160
  stages, 155
military, 160, 168
production, 142; inventor-doctors, 161, 163; motivations, 162–3
the promise of, 166–8
psychology, 154–9
semi-, 147–8
society, 141–4
technologies, 142; gendered, 160
wheelchair, 167
writing, 211

Dadaists' manifesto, 206
Dead/Death(s)
  and disability, fear of, 163; motivates funders, 152

Dead/Death(s) (*Continued*)
   as enemy, 162; defeating, postponing, 170n
   prolonging, 147–8
Debakey, Dr. Michael, 147, 153
Democracy
   and AT&T, 92
   and information, 73–4
   mythinformation, 73
Department of Energy (DOE), 33, 34, 48, 50, 60n, 61n, 62n
   military production, 35
   and New Mexico agreements, 52; disagreements, 41–8
   public relations, 40; labeled critics, 52; refuses dialogue, 54
   violations, 43
   and Westinghouse Corp., 53
DeVries, Dr. William, 151, 152; Symbion stockholder, 170n
DGA, *see* Directors Guild of America
Dialysis, 147, 166
Directors Guild of America (DGA), 5, 13, 16–9, 23n
   motivations, 6
   versus Motion Picture Association, 12
Directors Guild of Great Britain, 15; against colorization, 23n
Disability and death, fear of, 163
Disciplinarity, 253–5; *see also* Interdisciplinarity
Discipline(s)
   boundaries, vii
   history of, 254
   new, 253; proliferation of, 254
   nursing as, 72
Discourse(s) of
   cyberspace, 184; cyborg, 210
   matrix, 188
   medical, 210
   military, 210
   rational, 179
   rhetorical, 179; melodrama, 198
   science, 179, 184

   technology, technocultural, 179
Disney, Walt, 219
   animating space flight, 220
   see also Walt Disney Co.
Dog research on livers, 149; hearts, 163
Doll(s), 233–49, *243*
Domenici, Senator Pete, 42, 45

Economic(s)
   of artificial heart, 151–2
   desire for cyborgs, 159–66
   of the sign/signal, 123
Efficiency, defining, 76; defined, 119–20
Eisenhower administration, 35
Engels, Friedrich, 198, 200, 206
Engineering
   communication, 119, 135
   dangers, 1
   engineers as heroes, 91
   faith, 1
   genetic, 81, 180
Environmental Protection Agency (EPA), 33, 40, 43, 47
Envy, cyborg, 188; uterus, 188
EPA, *see* Environmental Protection Agency
Ethics
   of bridge transplants, 153
   of cyborgs, 163
   delays, 163
   research, 163
   and technology, vii
Externalist history of technology, 257

Faith, in technology, 93–5
Fascism, and futurism, 204
FDA, *see* Federal Drug Administration
Fear, of disability and death, 163
Federal Drug Administration (FDA), 152; toothless, 153
Feedback, 134, 168n
   bio, 151

Index

origins, 120
  in manifestos, 210
  rhetorical, 209–11
Feminism(s); femininity, 68
  manifestos, 206
  Marxist, 208
  Mina Loy's manifesto, 206
  radical, 208
  science and technology analysis, 159, 196, 206–9
Film, *see* Motion pictures
Film Disclosure Act, 19
Film Integrity Act, 16, 17
Final Environmental Impact Statement, WIPP, 43, 47
First World War, *see* War, First World
Fordism, 201; and Post-Fordism, 182
Foreign affairs and space policy, 220–2, 225
Foucault, Michel, 186, 192; terms, 197
Fourier analysis, 128, 200
Freedom of Information Act, 40
French and Indian War, *see* War, French and Indian
Frontier
  imagery, 228n
  metaphor and space exploration, 217, 222–4
Fulling, defined, 114n
Future
  transforming machines, organic, 212
  yours, xi
Futurism's new man, 196
Futurist manifestos, opposition to, 206

Garvey, Marcus, 236
Gender, *see also* Feminism(s); Masculine
  constructions, 210–1
  and cyborg, 159; dependence, 154
  defined, 68; in language, 205
  identity, 68
  ideology, 67; and masculine assumptions, 160
  and manifestos, 206
  and Marx, 212n
  and nursing, 65–79
  post, 208
  power, 77
  and technology, viii; relationship, 65–79
Genetic engineering, 81; manipulations, 142
Gephardt, Rep. Richard, 16–7
Gibson, William, 179–80, 192
  and the technological sublime, 213n
Gigi, 29
Glenn, John Jr., 223, 227n
Goffman, Erving, 120, 134, 135, 136n
Goldman, Emma, 94
Goodman, Paul, 2, 4
Greely, Horace, 217

Habermas, Jurgen, 179
Hal Roach Studios, 9, 31
Haraway, Donna, 142, 148, 159, 168, 174, 187–8, 192–3, 206–9, 214
  cyborg, 196; use of, 210
  and history, 212
  holism, 209
  manifesto, 214
  and technological sublime, 213n
Hay, John, 86, 112, 117n
H. B. Leonard Films, 31
Heart
  artificial, 150–2, 166–7
  bridge transplants, 153
  dog's pulsating, 163
  lung transplants, 169n
Heinlein, Robert, 162, 221, 229n
Hemingway, Ernest, 126, 132, 161
Hesse, Von Maxeda, 234, 235, 237, 239, 240

Heywood, Bill, 94
Hiroshima, 131, 188
History
   artificial kidney, 166
   demography, 66–7
   dolls, colored, 235–7
   of ideas, 253
   medicine, 160
   nursing, 66–7
   postmodern, 256
   public policy, 216
   of science, 253
   of Society for the History of Technology, 256–7
   technology, viii, 179, 251–3, 257
      approaches, 257
      and critical thinking, ix
      importance of, vii, x; why, 255
   telegraph operators, 138n
History of Science Society, 253
Hitler, Adolf, 215
Hospital Corporation of America, 170n
HR 2400, see Film Integrity Act
HR 3501, see Film Disclosure Act
Hubbard, Elbert, ix, 81
   background, 82–4
   death, 84, 94
   marketing, 82; defends trusts, 94
   The Philistine, 83, 92
   politics, 84, 94
   technology, 85
   youth, 82
Hughes, Rep. William, 18, 19
Humana Inc., 151–2; Symbion stockholder, 170n
Humans
   adaptable, 156; defined, 2
   ambiguity, source of, 120
   bodies, 167
   brain, 131
   as components, 129–32
   consciousness, 166
   cyborgs, division from, 131, 154–5, 166, 189, 209; integration, 148, 159, 165

downloading, 166
emancipation, 181
enhanced, 164, 165
functions, 131
as gods, 166
mechanization, 129–32, 131, 159, 167; symbiosis, 148, 165
   boundary, 131, 209
   compression, 189
   integration, attitudes toward, 159
   learning, 132
   networks, 212; systems, 207
   power, 212
   relationship, 142
physiology, 120
poor versus rich, 166
post, 164
purpose, 2, 215
survival, and space exploration, 215
transportation, 148
Hurston, Zora Neale, 243
   Sara Lee Doll, 237–8; encouraging, 239
Hussein, Saddam, 215
Huston, John, 16, 247n; colorization, 7, 12, 20
Hybrid
   as hybrid, 209–11
   organic, 209
   machine, 209
   text, 200
   see also Artificial organs; Livers
Hyper-reality, 185

Idaho Engineering Laboratory, 37
Ideal Toy Co., Sara Lee Doll, 234, 239
   contract, 146
   delays, 242
   publicity, 145–6
Ideologies of
   control, 120
   cyborg, 157
   gender, 72

Index

logic, 72
rehabilitation, 155–6
technology, 81, 159
Image Transform Laboratory, 7
Immortality, 164, 165; cyborg, 167
Immune system, as software, 148
Immunizations, 142, 169n
Infomedicine, *see* Medical/Medicine, info
Information
  accurate, 74
  age, 72, 75, 81
  assumptions, 73–4
  encoded, 122
  and the Grain Exchange, 122
  managing, manipulation, 73
  mythinformation, 73–4
  nursing, 72
  polymorphous, 207
  processing, 122
  science, 120, 169n; theory, 120; origins, 120; *see also* Cybernetics
  systems, 207; bedside, 76
  technology, *see* Technology, information
  theory, 120, 129–32, 136n
Interagency Review Group (IRG), 36, 38, 42
Interdisciplinarity, x, 254; *see also* Discipline(s)
Interior Department, 45, 50
Internalist, history of technology, 257
International HRS Industries, 9
International Photographers' Guild, 24n
International relations, and space exploration, 227
International Society for Artificial Organs (ISAO), 159, 168n; journal, 162
Interracial alliance, 234, 246
Inter-Racial Council, 234–5, 240; Creech in, 235

INTV, *see* Association of Independent Television Stations
ISAO, *see* International Society for Artificial Organs

Japan's artificial liver program, 149
Johnson, President Lyndon, 222
  letters, 232n
  and NASA, 225
Jones, Mother, 94

Karp, Haskell, transplant victim, 153; Mrs. Karp, 171n
Kennedy, President John, 216, 222, 231n
Kidney, artificial, 147–8; transplants, 169n
Knowledge, forms, 70; is power, 74
Kolff, Willem, 147, 150, 151, 154, 160, 161, 163, 170n, 175
Kolff Associates, *see* Symbion Inc.

Lake Sunapee, New Hampshire, 97, 108; changes, 109; recreational, 113
Language
  American, telegraphic, 123
  class, 137n
  common, 207
  compression, 123
  encryption, 123
  future, 123–4
  mechanization, 131
  natural, 131
  punctuation, 126
  standardization, 123, 125; statistical properties, 124
  a string of symbols, 120
  of systems, 207
  and telegraphs, 122–7
Lenin, 200, 201
Lexicon Corporation, time compression, 23n
Lexiconning, defined, 23n
Ley, Willy, 219, 230

Library of Congress, 16, 30; copyright office, 17, 26
Life prolonging, 147–8; support, 154
Livers, artificial, 147, 148–50; transplants, 169n; first, 169n
Logic
 of the excluded middle, 200
 as ideology, 72
 of manifestos, 203
Los Alamos National Laboratory, 33, 41, 48, 56

Machine/Machinery
 age, 252
 cognition, 129
 cyborg, 157
 dependent, 154–7
 human
  boundary, 131
  connections, 156; integration, 159, 165; relationship, 142
  dependence, 154
  learning, 132
 joy, versus logic, 203
 as metaphor, 252
 not an it, 168
 organic, 209; future of, 212
Magic, of technology, 191; cyberspace, 183
Male, see also Feminism(s); Gender; Masculine
 German attitudes toward war, 213n
 status and power, 72
Man, conforms, 3
Manifesto, x
 Communist, 196–202
 cyborg, 195, 206–9
 language, genre, 196; rhetoric, insults, 207
 monster, 195–6
 origins of, 196–7
 technologies, 195–214
 violence, 205

Man-machine weapon systems, 167
Marinetti
 deus ex machina, 202–4
 influence on D. H. Lawrence, 213n
 manifesto, 195–214
 violence and precision, 195–6
 and war, 209
Markets, perfection, 122; see also Motion pictures, markets
Marshall, Thurgood, 241
Marx, Karl
 and Engels, 198, 200, 206
 feminism, 208
 on gender, 208, 212n
 homo faber, 196
 manifestos, 195–214; Communist, gender, 212n
 rhetoric, 202; master narrative, 198
 scientific socialism, 199
 on war, 206
Masculine, fantasies, 188; qualities, 160; themes, 162
Mass media, see Media, mass
Mathematical, abstraction and communication, 119–20, 127–9
McLuhan, Marshall, ideology of technology, 81
Media, mass, 5, 142; analyzing, 95
Medical/Medicine, see also Nurses
 bio-, 163; discourse, 164
 care and class, 166
 committees, 161
 cyborg, 144, 210–1
 discourse and cyborg, 164, 210–1
 Galenic, 163, 164
 gender, 160
 info-, discourse, 164
 malpractice, 171n
 masculine and feminine, 160
 science and values, 164

Index

surgeon's interventions, 147
Melodrama, discourses, 198; postmodern by Haraway, 208
Metaphors, of cyberspace, 180, 187; magical, organic, nautical, 183–5,
MGM-UA, colorization, 10, 14, 31
Military
  and cyberspace technology, 181
  cyborgs, 160; discourse, 210–1
  power and space exploration, 221–2
Motion pictures, 5
  altered, 18; versus artistic, 19
  classics, 15, 18, 19
  colorized effects, 17; ratings, 11
  copyright holders listed, 31
  distribution windows, 12
  markets, 13
  national treasures, 28–30
  technology, 5
  television, 12; triumph of, 13
Motion Picture Association of America (MPAA), 6, 13, 19, 24n
  versus Directors Guild, 12
  massive lobbying, 17
Movies, *see* Motion pictures
MPAA, *see* Motion Picture Association of America
Mrazek, Rep. Robert, 17, 19, 21; colorization, 18; amendment, 17
Multidisciplinarity, 254; *see also* Discipline(s)
Mythinformation, 73
Myths, Greek, 144

NAACP Legal Defense Fund, 240–1
NAB, *see* National Association of Broadcasters
Nader, Ralph, 36
Nanotechnology, 179–80
Narrative of cyberspace, 186
NASA (National Air and Space Administration), x
  and European Space Agency, 226
  and frontier myth, 223
  History Office, xiii, xvi
  and Russians in space station, 226
  space policy, 215–32
  and virtual reality, 180
National Academy of Sciences, 27, 49
National Air and Space Administration, *see* NASA
National Association of Broadcasters (NAB), 13, 24n
National Association of Theatre Owners, 24n
National Endowment for the Humanities (NEH), xiii, 168n
National Environmental Policy Act, 36, 43
National Film Preservation Act, 15, 17, 18, 21, 22, 27, 28; toothless law, 19
National Film Preservation Board, composition, 23n
National Institutes of Health, 151
  Artificial Heart Assessment Panel, 170n
  bureaucrats and cyborgs, 159
  funding limits, 171n
  regulation violations, 152
  Working Group on Mechanical Circulatory Support, 166
National security and space policy, 220–2, 225–6
National Society of Film Critics, 23n
Negro, *see* African-American
NEH, *see* National Endowment for the Humanities
Nevada, rejected Yucca Flats, 50
New Hampshire, 97–118; *see also* Lake Sunapee; Newport
  history, 98
  ideal community, 98
New Mexicans for Jobs and Energy, 50

New Mexico
  Attorney General Tom Udall, 46
  and Department of Energy, 41–8;
    agreements, 52; financial, 46
  Energy Research and Development Program, 47
  Health and Environment Department, 53
  land commission, 44
  licensing review, 43
  officials, 35; governor's office, 43; legislature, 42; senators, 45
  see also Carlsbad; Department of Energy; Waste Isolation Pilot Project
Newport, New Hampshire, *106*
  1800s, 101; growth, 98–9
  mid-1800s growth, 105
  1840 census, 104
  dam, 108; see also Lake Sunapee
  economic base, 104, 113
  ethnic composition, 117n
  map, *107*
  settlement, 97
Nixon, President Richard, 232n; and NASA, 225, 231n
Nosé, Dr. Yukihiko, 143, 163, 176
Nuclear energy
  powered artificial hearts, 170n
  program, 45
  public perceptions, 35
Nuclear Regulatory Commission (NRC), 36, 38
Nuclear Waste Policy Act, 40, 45
Nuclear wastes
  disposal, 37
  and weapons, 57
Nuclear weapons, and waste, 57
Nurses/Nursing
  art, 73; or science, 72; unquantifiable aspects, 77
  automation as enhancement, 75; history of, 70
  caring as essence, 74
  class, 67

computerization, 70, 73–7; overpromised, 75
costs, 74
demography, 66–7; supply, 76
economic worth, 75
efficiency, 77; defined, 76
gendered, 72; ideology, 67–8; demographics, 66; and technology, 65–74
history, 66–7, 70; in literature, 65
practice, and technology, 68–73; productivity, 71; quality, 76
race, 66
station, invented, 71
status seeking, 73
stress, 72
Taylorism, 70
technology, and gender, 65–74
a women's profession, 67–8

O'Leary, Hazel, secretary of energy, 33
Organic
  cyberspace metaphor, 183
  hybrid, 209
  world view, 209
Organs, see also Artificial organs
  animal, 148; baboon, 169n
  artificial, 141; penis, 146
  interchangeable, 164
Owen, Donald, 200; Symbion stockholder, 170n
Oz, 83; like cyberspace, 183

Penis, artificial, 169n
  Pearman's, 146
  Scott's inflatable, 146
  supplemental, 146, 158
Persian Gulf War, see War, Persian Gulf
The Philistine, 82, 84–9; founding, 83
  ads, 88; cigarette banned, 86
  and American big business, 93
  and AT&T, 89

Index

competition, 87
Philosophy, moral, technology as, 2
Politics
　artifacts, 212
　of literary productions, 197
　technocratic, 190
Post-Fordism and Fordism, 182
Posthuman, 141–78
Postmodern, melodrama, Haraway's, 208
　history, 256
　medicine, 164
Power
　gender and nursing, 67–8
　knowledge, 181
　rhetorical strategies, 184
　of speed, 208
　systems of human-machines, 212
　texts of cyberspace, 180
Presidents, *see* Bush; Carter; Johnson, Kennedy, Nixon, Reagan
President's Science Advisory Committee, 224, 228n
Pressure and force, defined, 115n
Private property, American allegiance to, 7
Profits, corporate, 167–8; medical, 160
Progress, inevitability, 89
Project Gnome, 36
Project Plowshare, 36
Prosthesis/Prosthetics
　amputee researchers, 163
　dreams, 158
　history, 171n
　hook, 145
　improvement, 145
　limbs and genitalia, 144–6
　medical, 165; bio, 141
　powered, 145
Psychoanalysis, reinvigoration, 192n; reconsideration, 186
Psychology
　of cyberspace, 185–6
　cyborg, 154–9; desire for, 159–66

medicine, cyborg, 163
Public historians and policy makers, 216
Public Law 96–164, 38
Public opposition to radioactive waste, 53
Public policy
　absence of, 39
　European radioactive waste, 39
　and historians, 216
　land and radioactive waste, 44
　space, x
　and technology, vii

Quadriplegics, 145
Quantification, 71; mania, 164
Queen Victoria, 86

Race, *see also* Interracial alliance; Inter-Racial Council
　discrimination campaign, 240
　history and technology, 233–49
　prejudice, and toys, 235; resisting, 236–7
Radiation/Radioactive
　dangers, 56
　for transplants, 169n; artificial hearts, 150
Radioactive waste
　contamination, 52
　disposal study, 37
　federal agency needed, 58
　funding, 38
　policy, 39; European, 39; Netherlands, 61n, 62n
　and potash industry, 55
　problems, 24; contamination, 52; risk, 58
　repository, first, 37; Yucca Mt., 45
　salt beds, 37; in Kansas, 37
　transport, 53
Rational(ity)
　devices, 191
　discourse, 179
　refocused in cyberspace, 181

Reagan, President Ronald, 17, 45
  administration, 42
  era, 36
  and NASA, 226
Rehabilitation, ideology, 155–6;
  and machines, 155
Representation, 181; genres, 212; of
  technology, 195
Republic Pictures, 31; colorization,
  11, 14
Residuals, defined, 23n; payments,
  13
Revolution(ary), 2
  biomedical, 151
  communication, implications,
    81
  control, 121
  industrial, 201, 252
  modernist, 182
  resistance, 201
  scientific, 2
Rhetoric
  discourse is, 179
  historical strategies, 180, 184
  Marxist, 202
  power, 184
Right(s)
  creative, 12, 16
  of disabled in Canada, 157
  to minerals, oil, gas, 44
  to private property, 7, 16; and
    copyright, 16
  for unaltered movies, 12
Risk
  acceptable, 40; with WIPP, 58
  calculating, radioactive waste, 53
  management, manipulated, 51,
    58; discussion, 61n
Rocky Flats Nuclear Facility, 37
Rogers, Ginger, 16
Roosevelt, Eleanor 234, 237,
    239–40, *243*
  saves Sara Lee Doll, 243; tea 246

Saddam Hussein, 215
SAG, *see* Screen Actors Guild

Sandew, Barry, colorization inventor, 10, 11
Sandia National Laboratory, 40, 41,
  47, 56, 61n; WIPP advisor, 49
Satellite
  artificial, advocated, 219
  Sputnik launching, 219–20
  Soviet versus American, 222
Science/Scientific
  and business, 84, 89
  community, 184
  discourse, 179; theories of com-
    munication, 129
  evidence, 71
  fiction, 179–210
    and the artificial heart, 150
    and cyborg medicine, 162
    described, 179–80
    hard and soft, 254
    limits, 187
    movies and space exploration,
      229n
    and space exploration, 217–8
  information, 169n; *see also* Cy-
    bernetics
  management, reforms, 75
  marketing, 89
  medical values, 164
  nursing, or art, 72
  postmodern, 164
  power, 166
  problems, 35
  revolution, 2
  style, 199
Scott's inflatable prosthesis, 146
Screen Actors Guild (SAG), 13, 24n
Self/Selves
  decentered, 186
  fragmented, 186
  fulfillment, 135
  identity, 124; and communica-
    tion, 132–5
  interest, of cyborg technicians,
    162
  and modernity, 181
  saturated, 186

Index

socially constituted, 135
Senate, U.S.
  judiciary subcommittee on colorization, 6
  Subcommittee on Technology and Law, 6, 14, 16
Sex, teledildonics, 181
Shelley, Mary, 130
SHOT, *see* Society for the History of Technology
Society
  of cyborgs, 141
  happy, defined, 98
Society for the History of Technology (SHOT), xi, 253, 256; history of, 256–7
Southwest Research and Information Center (SRIC), 50
Space
  commercial exploitation, 224
  Euclidian, 182
  exploration, 1, 217–20
    American, 218; 1950s, 216–22; 1990s, 222–6; new, 225–6
    costs, 225
    and the frontier, 216, 217
    and international relations, 220–2, 227
    military power, 221–2
    motivation, 215
    policy formation, 227; role of analogy, 215
    race, 225
    reality, 224–5,
    and science fiction, 217, 229n
  as a fault of nature, 181
  Shuttle, 224–5; *Challenger* accident, 225
  station, costs, 227; *Freedom,* 226; with Russians, 226
  and time, 179, 180, 181; relative and compressed, 182
Spanish-American War, *see* War, Spanish-American
SRIC, *see* Southwest Research and Information Center

Stewart, Jimmy, and colorization, 7, 12, 14, 15, 18; goes to Washington, 15–9
Symbion Inc., artificial hearts, 171n; stockholders, 170n
Symbiosis, between life and culture, 252; in cyberspace, 189
Symbolist, aesthetic, 203
Symbols
  Egyptian and AT&T, 93
  language, a string of, 120
Systems
  computer and human nervous, 130, 207; human-machine, 144
  cybernetic, 168–9n
  language, 207
  Nile and the phone system, 93

Taylor, Frederick, and Taylorism, 70–1, 127
Technical level, 49
Technicians, hazardous materials, 48
Technobabble, 256
Technoculture, 256
Technohistorians, 195; histories, 196
Technohistory, xi, 255–8
  defined, 256
  implications, 257
  readings, 259–62
  teaching critical thinking, 81–95
  technologists, need technohistory, 256
Technology, defined, 70
  accidents, 45; *Challenger,* 225; acceptable risk, 40; failures, 244–5
  age of, 66
  of agency, 212; constructed, 209
  alienation, 36
  and Americans, 82, 99
  assumptions, 81; integral, x
  as black box, 3

Technology (*Continued*)
  of the body, 182; grounding in body, 68
  and civilization, 251
  and class, 209
  communication mathematicized, 129
  control through communication, 121
  discourses, high, 179; word defined, 2; writing, 195, 211
  ethics, violations, 152
  faith in, 1, 93–5; fascination, 190
  gender, 65; and nursing, 65–79
  grounding, 68
  history of, 99, 151–3, 179, 255, *see also* History, technology
  as human behavior, 2, 3
  ideology, 81, 159; new, 202–3
  and information, 73; in hospitals, 69, 73–77
  of magic, 191
  of management, 121
  manifestos, 195–214; and politics, 206
  military, cyberspace, 181, 183
  as moral philosophy, 2
  mythinformation, 73
  neutral, not, 70
  optimism, viii, 36; age of, 59, 60n; opportunity, 167; promise, 76
  problems with, 35; waste storage, 52; opposition, 48
  representation of, 195, 212
  sublime, 13n
  transparent, 253
  and war, 188; violent, 201; cyberspace, 181, 183
  and women's lives, 49, 89, 253; gender, 65–79
Technophilia, 204
Technopolis, 256
Telegraph(y)
  acoustic, 128
  information processing, 122
  and language, 122–7; compressed, 119; punctuation, 126
  malfunctions, 125
  150 year anniversary, 119
  style, 123
  and the telephone, 132
Telephone, 133
  as annihilator of space, 92
  business market, 132
  fees, 133
  first 50 years, 132
  long-distance, 128
  mapping the social world, 134
  and telegraph, speaking, 132
  and women, 133
Telepresence, 181
Television, motion pictures altered for, 12, 13
Test(s), nuclear
  above-ground, 36
  WIPP, 47, 56; canceled, 46
Text(s)
  cyberspace, 186
  false or suspect, 212
  hybrid, 200
TIAH (Totally Implantable Artificial Heart), 150
Time and space categories, 179, 181; changes, 180; relative, 182
Tintoretto, Inc., 10, 14; inactive, 23n
TNT, 13
Toys, industry, 234; as agents of social change, 245
Transdisciplinary, 254; *see also* Discipline(s)
Transplants
  history, 148, 169n
  number of, 169n
  successful, 169n
Turner, Ted
  aspiring villain, 7
  buys MGM library, 13
  dead issue, 20; prediction of, 22–3
  evil genius, viii

Index 279

Mouth of the South, 6
surrogate Philistine, 20
Turner Entertainment Co. 11, 14, 20, 31; artistic desecration, 16
TV, *see* Television
20th Century-Fox, 31; colorization, 10, 11

Union of Concerned Scientists (UCS), 61n
United States, *see* America(n); Congress; Environmental Protection Agency; Federal Drug Administration; Library of Congress; Senate
U.S. District Court, 43
U.S. government versus the states, 42
  New Mexico, WIPP, 46
Universes, Euclidian, Newtonian, Einsteinian, 182
University of Kansas, waste report, 37
University of Utah, 151–2; Institutional Research Board, 152
Urban League, 239; in St. Louis, 237
Uterus envy, 188

VD, *see* Video Software Dealers of America
Video recorder market penetration, 13
Video Software Dealers of America (VD), 13
Videotapes, colorized, 13
Violence, in manifestos, 206
Virgin Mary, 68
Virtual Reality (VR), 180; worlds, 189
VR, *see* Virtual reality

Walt Disney Co., 31
War
  armies as first machines, 252

automation, 138n
Cold, 25, 221; rivalries, 222; end of, 45
class, 206
cyberspace, 181
First World, 93; and the mechanized body, 213n
French and Indian, 97
futurism, glorifying, 204
German, male fantasies, 213n
Marinetti, glorifying, 209
Marx's notion, 207
metaphor, 206
nuclear policy, 221; opposition to, 161
Persian Gulf, 181
Second World, 129; impact on toy industry, 236
space policy, 221
Spanish-American, 87
technology, cyberspace, 183; uterus envy, 188
Washington District Court, New Mexico sues WIPP, 46
Waste Isolation Pilot Project, 33, 35–9; *see also* Radioactive waste; Test(s)
  accidents, potential, 47, 53, 58; radiation, 56
  alternate sites, 47
  authorized, 38, 41
  Carlsbad, N.M., 55–7; *see also* Carlsbad, New Mexico
  compensation figures, 62n
  engineering feat, 54
  exempted from regulations, 52; illegal land transfer, 44, 46
  funds, 47, 55
  future, 57
    cancellation, 59
    opening, 58–9
    postponing, 58–9
    suspended, 39
  jurisdiction, 48
  military wastes, 34
  new shaft, 50

Waste Isolation Pilot Project (*Continued*)
  policy, 61n; inadequate, 35
  problems, 57, 58
    accidents, 47
    leaks, 56
    risks, 53
    technical, 35, 46, 52, 58
  public opposition, 34, 48–55, 59
    first public protest, 49
    hearings, 54–5; first, 44
    injunction, 45
    labeled, 52
    polls, 48
    referendum, 48
    strategy, 51; emotional appeals, 54
  risks, 47, 53, 58; radiation leakage, 52, 56
  salt compression, 56; brine, 52
  scientific and technical problems, 35
  Scientists' Review Panel, 52
  states' role, 42
  testing, 56; planned, 47; canceled, 46; injunction, 35
  transport, risks, 53
  vertical migration, radiation danger, 56
Waste Policy Act, 59
wastes, non-radioactive, 38
wastes; radioactive, transporting, 46; Yucca Flats rejected, 50–1
Waterwheels, change, 102; reaction, 103; overshot, undershot, 99–100
Western Union, 127; versus AT&T, 129
Westinghouse Corporation, and Department of Energy, 53; WIPP, 41
Wheels, *see* Waterwheels
Wiener, Norbert, 120, 130, 136n, 138n
WIPP, *see* Waste Isolation Pilot Project
Women
  ideal, 65
  scorn for in futurism, 204
  and technology, 65
  and telephones, frivolous use of, 133
Writers Guild of America (WGA), 13, 23n
Writing, technologies, 195, 211
WTBS, 13

xenotransplants, 148–9, 169n; baboon, 170n